NEW CENTURY MATHS

MATHEMATICS STANDARD 1

YEAR 12

Sue Thomson
Judy Binns
Series editor: Robert Yen
2ND EDITION

New Century Maths 12 Standard 1
2nd Edition
Sue Thomson
Judy Binns

Publisher: Robert Yen
Project editors: Alan Stewart and Anna Pang
Editor: Lisa Schmidt
Cover design: Chris Starr (MakeWork)
Text design: Sarah Anderson
Art direction: Aisling Gallagher and Danielle Maccarone
Cover image: iStock.com/OGphoto
Permissions researcher: Helen Mammides
Production controller: Christine Fotis
Typeset by: Cenveo Publisher Services

Any URLs contained in this publication were checked for
currency during the production process. Note, however, that
the publisher cannot vouch for the ongoing currency of URLs.

For product information and technology assistance,
in Australia call **1300 790 853**;
in New Zealand call **0800 449 725**

For permission to use material from this text or product, please email
aust.permissions@cengage.com

National Library of Australia Cataloguing-in-Publication Data
A catalogue record for this book is available from the National Library of
Australia.

Cengage Learning Australia
Level 7, 80 Dorcas Street
South Melbourne, Victoria Australia 3205

Cengage Learning New Zealand
Unit 4B Rosedale Office Park
331 Rosedale Road, Albany, North Shore 0632, NZ

For learning solutions, visit **cengage.com.au**

Printed in China by China Translation & Printing Services.
2 3 4 5 6 7 22 21 20 19 18

CONTENTS

*Year 11 revision

1

2

3

4

10

HEALTHY HEART   286

11

APPLYING TRIGONOMETRY   308

12

FROM PAPER TO REALITY   346

13

UNBIASED DATA   390

PREFACE

New Century Maths 12 Mathematics Standard 1 has been rewritten for the new Mathematics Standard syllabus (2017). In this 2nd edition, teachers will find those familiar features that have made *New Century Maths* a leading mathematics series, such as clear worked examples, graded exercises, syllabus references, investigations, language activities, chapter summary mind maps, practice exercises and a glossary/index.

Mathematics Standard 1 is designed for students who require a broad knowledge and understanding of mathematics to apply in employment and everyday life. We have endeavoured to produce a practical text that captures the spirit of the course, providing relevant and meaningful examples of mathematics being used in society and industry.

The *NelsonNet* student and teacher websites contain additional resources such as worksheets, video tutorials and topic tests. We wish all teachers and students using this book every success in embracing the new senior mathematics course.

ABOUT THE AUTHORS

Sue Thomson was head of mathematics at De La Salle Senior College, Cronulla, Director of Teaching and Learning at Hunter Valley Grammar School, and an HSC examination writer, assessor and marker. Sue's interests are in language development, financial literacy and making mathematics accessible to all.

Sue dedicates this book to the memory of her husband and co-author, **Ian Forster**.

Judy Binns was head teacher of mathematics at Mulwaree High School in Goulburn and has taught at Homebush Boys High School. She has wide experience teaching and writing for practical senior mathematics courses. Judy has been co-writing the *New Century Maths 7–8* series for over 20 years.

Series editor **Robert Yen** has been writing for *New Century Maths* since 1995, as well as writing and presenting for MANSW and co-editing its journal, *Reflections*. Robert now works for Nelson Cengage as the mathematics publisher.

CONTRIBUTING AUTHORS

Deborah Van Hoek (Westfields Sports HS) and **Kuldip Khehra** (Quakers Hill HS) wrote and edited many of the *NelsonNet* worksheets.

John Drake, Katie Jackson and **Joanne Magner** created the video tutorials.

ISBN 9780170413534

SYLLABUS REFERENCE GRID

Topic and subtopic	New Century Maths 12 Mathematics Standard 1 chapter
Algebra	
MS-A3 Types of relationships	3 Graphing lines
A3.1 Simultaneous linear equations	6 Graphing curves
A3.2 Graphs of practical situations	
Measurement	
MS-M3 Right-angled triangles	9 So you've got a right angle
	11 Applying trigonometry
MS-M4 Rates	7 Applying rates
	10 Healthy heart
	12 From paper to reality
MS-M5 Scale drawings	2 What's my share?
	12 From paper to reality
Financial Mathematics	
MS-F2 Interest	1 Investing money
MS-F3 Depreciation and loans	4 Depreciation and loans
Statistical Analysis	
MS-S3 Further statistical analysis	5 Fitting the data
S3.1 The statistical investigation process for a survey	13 Unbiased data
S3.2 Exploring and describing data arising from two quantitative variables	
Networks	
MS-N1 Networks and paths	8 Finding the right path

ABOUT THIS BOOK

AT THE BEGINNING OF EACH CHAPTER

- Each chapter begins on a double-page spread introducing the **Chapter problem**, outlining the **Chapter contents** with syllabus codes, a list of chapter outcomes titled **What will we do in this chapter?**, and a list of applications titled **How are we ever going to use this?**

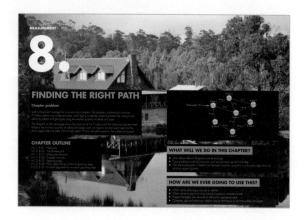

IN EACH CHAPTER

- Important facts and formulas are highlighted in a shaded box.

- Important words and phrases are printed in red and listed in the glossary at the back of the book.

- Graded exercises are linked to worked examples and include exam-style problems and realistic applications.

- **Investigations** and **Practical activities** explore the syllabus in more detail, providing ideas for modelling activities and assessment tasks.

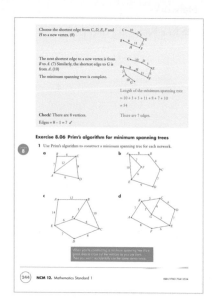

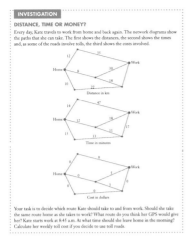

DISTANCE, TIME OR MONEY?

Every day, Kate travels to work from home and back again. The network diagrams show the paths that she can take. The first shows the distances, the second shows the times and, as some of the roads involve tolls, the third shows the costs involved.

Your task is to decide which route Kate should take to and from work. Should she take the same route home as she takes to work? What route do you think her GPS would give her? Kate starts work at 8:45 a.m. At what time should she leave home in the morning? Calculate her weekly toll cost if you decide to use toll roads.

d Mandy doesn't feel very well. She is finding it hard to concentrate and she has a headache. When she took her blood pressure on her home monitor, it was 195/108. What do you recommend Mandy do?

4 This table shows normal, resting blood pressure readings for children.

Age	Systolic blood pressure	Diastolic blood pressure
Birth (12 hours)	60–76	31–45
Neonate (96 hours)	67–84	35–53
Infant (1–12 months)	72–104	37–56
Toddler (1–2 years)	86–106	42–63
Preschooler (3–5 years)	89–112	46–72
School age (6–9 years)	97–115	57–76
Preadolescent (10–11 years)	102–120	61–80
Adolescent (12–15 years)	110–131	64–83

Source: eMedicineHealth.com

a Rachel is 8 years old and her blood pressure is 98/60. Does she have normal blood pressure?

b Approximately how old is a healthy child whose blood pressure is 90/70?

c By approximately what percentage does a baby's blood pressure increase from when it is 12 hours to when it is 96 hours old?

d Describe how a child's blood pressure changes with age.

PRACTICAL ACTIVITY

BLOOD PRESSURE AFTER EXERCISE

For this group activity, you need an automatic blood pressure-measuring machine and access to a walking area.

1 Search the **Mayo Clinic** website or YouTube for a video 'How to measure blood pressure using an automatic monitor.'

2 Measure each group member's resting blood pressure.

3 Go for a brisk group walk for 5 minutes, then measure everyone's blood pressure again.

4 Construct a scatterplot showing each person's resting blood pressure against their blood pressure after their walk.

5 What happens to blood pressure after exercise?

6 Does exercise change everybody's blood pressure in the same way?

7 Describe the resting blood pressure of individuals whose blood pressure changed the most during exercise.

AT THE END OF EACH CHAPTER

- **Keyword activity** focuses on the mathematical language and terminology learned in the chapter.

- **Solution to the Chapter problem** revisits the problem introduced at the start of the chapter and presents a solution to the problem.

- **Chapter summary** concludes the chapter and includes a mind map exercise.

- **Test yourself** contains chapter revision questions linked to the relevant exercise set.

- **Practice sets** revise the skills and knowledge of previous chapters.

KEYWORD ACTIVITY

SIMILAR WORDS

In this chapter, there are several pairs of words that have a similar but slightly different meaning. Explain, in the context of this chapter, what each word in the pair has in common and how they are different.

1 Vertex – vertices

2 Edges – weighted edges

3 Directed network – undirected network

4 Spanning tree – minimum spanning tree

5 Prim's algorithm – Kruskal's algorithm

SOLUTION TO THE CHAPTER PROBLEM

Problem

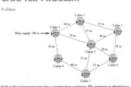

Isiah is the project manager for a construction company. His company is planning to construct 7 holiday cabins in a wilderness area. Isiah's task is to decide where to position the underground electricity cables and gas pipes using the smallest quantity of cables and pipes.

The diagram above shows the positions of the 7 cabins and the distances between them. What is the minimum quantity of cables and pipes Isiah will need to connect each cabin to the main supply that is located 100 m from cabin 1? How should he position the cables and pipes?

Solution

Use Prim's or Kruskal's algorithm to construct the minimum spanning tree.

Total amount of cable and pipes = 100 + 50 + 40 + 25 + 30 + 20 + 25
= 290 m

Isiah will require 290 m each of electricity cable and gas pipes.

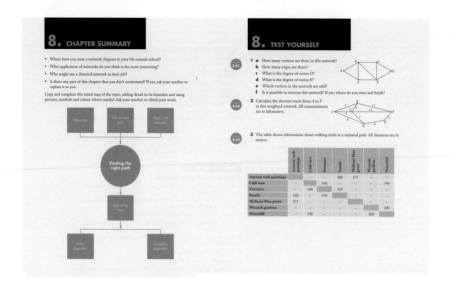

AT THE END OF THE BOOK

- **Glossary and index** includes a comprehensive dictionary of course terminology.
- **Answers**.

NELSONNET TEACHER WEBSITE

Margin icons link to print (PDF) and multimedia resources found on the *NelsonNet* teacher website, **www.nelsonnet.com.au**. These include:

Worksheets   Puzzle sheets   Skillsheets   Video tutorials   Spreadsheets   Weblinks

- **Worksheets** and **puzzle sheets** that are write-in enabled PDFs
- **Skillsheets** of examples and exercises of prerequisite skills and knowledge
- **Video tutorials**: worked examples explained by 'flipped classroom' teachers
- **Spreadsheets**: *Excel* files
- **Weblinks**
- A **teaching program**, in Microsoft Word and PDF formats
- **Chapter PDFs** of the textbook
- **Resource Finder**: search engine for *NelsonNet* resources.

> Note: Complimentary access to these resources is only available to teachers who use this book as a core educational resource in their classroom. Contact your Cengage Education Consultant for information about access codes and conditions.

NEW CENTURY MATHS AND MATHS IN FOCUS 11–12 SERIES

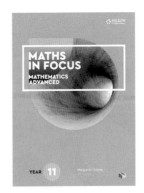

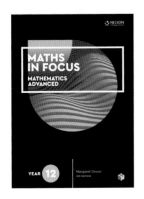

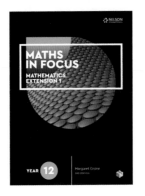

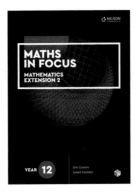

ISBN 9780170413534

FINANCIAL MATHEMATICS

1

INVESTING MONEY

Chapter problem

Ryan invested $25 000 for 3 years, earning compound interest at 4% per year.

Melanie also invested $25 000 for 3 years, earning simple interest at 4.2% per year.

Whose investment earns more interest?

CHAPTER OUTLINE

Note: Simple and compound interest graphs are covered in chapters 3 and 6 respectively.

WHAT WILL WE DO IN THIS CHAPTER?

- Revise percentage calculations, profit and loss, and simple interest
- Solve problems involving compound interest and inflation using the compound interest (future value) formula
- Compare simple interest with compound interest investments
- Use online calculators for interest calculations
- Recognise and avoid potential financial scams

HOW ARE WE EVER GOING TO USE THIS?

- When making financial decisions involving buying, selling and investing
- When checking the growth of an investment or an increase in prices
- To avoid falling for financial scams

1.01 Percentages and money

Calculating percentages accurately is an essential part of financial management.
Let's review 3 types of percentage calculations:

- Calculating a percentage of an amount (Example 1)
- Increasing or decreasing a quantity by a percentage (Example 2)
- Calculating the percentage represented by an amount (Example 3)

Percentage
power

Percentage
shortcuts

Percentages
without calculators

Working with
percentages

Percentage
problems

Percentage
calculations

Mental
percentages

Simplifying
fractions

EXAMPLE 1

Ben earns 5% commission on all sales he makes. Calculate Ben's commission for selling a car for \$14 990.

Solution

Write 5% as either $\dfrac{5}{100}$ or 0.05 and multiply.

$$\text{Ben's commission} = 5\% \times \$14\,990$$
$$= \$749.50$$

EXAMPLE 2

Bianca works in a clothing store. As part of her job, she adds GST by increasing all prices by 10%. Calculate the price of a pair of jeans priced at \$60 after GST is added.

Solution

There are two ways to do this. Choose the method you like!

Calculate 10% of \$60, then add to \$60.

$$\text{GST} = 10\% \times \$60$$
$$= \$6$$
$$\text{Price including GST} = \$60 + \$6$$
$$= \$66$$

OR The new price will be
100% + 10% = 110% of \$60.
Do $\dfrac{110}{100} \times 60$ or 1.10×60.

$$\text{Price including GST} = 110\% \times \$60$$
$$= \$66$$

ISBN 9780170413534

Percentage profit and loss

$$\text{Percentage profit} = \frac{\text{profit}}{\text{cost price}} \times 100\%$$

$$\text{Percentage loss} = \frac{\text{loss}}{\text{cost price}} \times 100\%$$

EXAMPLE 3

Rachel paid $650 000 for an apartment and she sold it for $790 000. Calculate Rachel's percentage profit correct to one decimal place.

Solution

Profit = selling price – cost price	Profit = $790 000 – $650 000
	= $140 000
Write the profit over the cost price and multiply by 100%.	$\text{Percentage profit} = \dfrac{140\ 000}{650\ 000} \times 100\%$
	$= 21.538 \ldots$
	$\approx 21.5\%$
Write the answer.	Rachel made 21.5% profit.

Exercise 1.01 Percentages and money

1 Calculate each amount correct to the nearest cent.

    a 8% of $75   **b** 9% of $135   **c** 2.5% of $360

    d 4.25% of $900   **e** 10% of $695   **f** 12% of $36.50

    g 15% of $27.95   **h** 11.5% of $129

Example 1

2 **a** Increase $720 by 5%   **b** Decrease $8600 by 9%

    c Increase $24 by 15%   **d** Decrease $746 by 6.5%

Example 2

3 Without GST, a pair of shoes costs $74. Calculate the price of the shoes with 10% GST added.

4 Lisa paid $550 for some shares that she later sold for $625. Calculate Lisa's percentage profit correct to one decimal place.

Example 3

5 The price of a new car increased from $15 000 to $16 500. Calculate the percentage increase in the price of the car correct to one decimal place.

6 Denis earns $32.60 per hour. Calculate his new wage per hour after he received a 2% wage increase.

7 On Saturday, the price of 1 L of petrol was $1.36 and on Monday it was $1.53. Calculate the percentage increase in the price of petrol.

8 Basam paid $198 for some concert tickets. The ticket agent's fee was 4.5% of the price of the tickets. Calculate the ticket agent's fee for the tickets.

9 Gretel's property was valued at $650 000 until the local council moved the 100-year flood line. Now, the property is worth only $425 000. Calculate the percentage decrease in the value of Gretel's property, correct to the nearest percentage.

10 Concession card holders pay only $2.50 for a return train trip to the city. The full return price is $19.60. What percentage of the full price do concession card holders pay? Answer correct to one decimal place.

11 During a sale, the price of a $2250 holiday was reduced by 7.5%. Calculate the sale price of the holiday. Answer to the nearest dollar.

12 In 2017, the Australian adult minimum wage for a full-time employee was $17.70 per hour.
   a The minimum casual rate was 25% more than the full-time rate.
   What was the minimum casual rate?
   b In 2017, the minimum wage for a 16-year-old full-time employee was $9.41 per hour. What percentage of the adult wage was the 16-year-old's wage? Answer correct to the nearest percentage.

1.02 Simple interest

When we invest money with **simple interest** or **flat-rate interest**, the amount of interest earned is the same every year. The interest is a percentage of the **principal**, the original amount invested.

Simple interest

What's the interest?

Simple interest riddle

Applications of simple interest

Simple interest

> ## The simple interest formula
> $$I = Prn$$
>
> where I = interest earned
>
> P = principal value invested
>
> r = rate of interest per period, expressed as a decimal
>
> n = number of periods

NCM 12. Mathematics Standard 1

ISBN 9780170413534

EXAMPLE 4

Chloe invested $6000 at 3.25% p.a. simple interest for 4 years.

| p.a. = per annum = per year |

a How much interest will she earn?

b How much will be in her account at the end of 4 years?

Solution

a In the simple interest formula $I = Prn$,
$P = 6000$, $r = 3.25\% = 0.0325$ and $n = 4$

$$I = 6000 \times 0.0325 \times 4$$
$$= 780$$

Write the answer.

Chloe will earn $780 in simple interest.

b Add the interest to the principal.

Account total $= \$6000 + \$780$
$= \$6780$

Write the answer.

Chloe will have $6780 in her account.

EXAMPLE 5

Roseanne invested $9500 at 4.2% p.a. for 30 months. How much interest did she earn?

Solution

The interest rate is per year, but the time is in *months*. We must change 30 months to years by dividing by 12.

$$\text{Time} = \frac{30}{12} \text{ years}$$

$P = 9500$, $r = 0.042$, $n = \dfrac{30}{12}$

$$I = 9500 \times 0.042 \times \frac{30}{12}$$
$$= 997.5$$

Write the answer.

Roseanne earned $997.50 in interest.

EXAMPLE 6

What percentage interest rate per month is equivalent to 5.52% p.a.?

Solution

Divide an annual interest rate by 12 to change it to a monthly interest rate.

$5.52\% \div 12 = 0.46\%$

Write the answer.

5.52% p.a. is equivalent to 0.46% per month.

EXAMPLE 7

Ranjan earned $420 in simple interest when he invested $6000 for 2 years. What was the rate of interest?

Solution

$I = 420, P = 6000, n = 2$

$I = Prn$
$420 = 6000 \times r \times 2$
$420 = 12\,000 \times r$

Solve the equation.

$420 \div 12\,000 = r$
$r = 0.035$
$r = 3.5\%$

Write the answer.

The rate of interest was 3.5% p.a.

Exercise 1.02 Simple interest

1 Calculate the simple interest on each principal.

   a $900 at 3.2% p.a. for 4 years

   b $1560 at 3.9% p.a. for 4 years

   c $4500 at 5.5% p.a. for $3\frac{1}{2}$ years

   d $2750 at 5.1% p.a. for $2\frac{1}{2}$ years

2 What is the simple interest on each investment?

   a $840 at 3.5% p.a. for 18 months

   b $12\,800 at 6.05% p.a. for 3 months

   c $2960 at $5\frac{1}{2}$% p.a. for 4 months

   d $880 at $6\frac{1}{4}$% p.a. for 1 month

3 What percentage interest rate per month is equivalent to 4.44% p.a.?

NCM 12. Mathematics Standard 1

ISBN 9780170413534

4 NCM Credit Union offers investors 7.2% p.a. simple interest.
Express this rate of interest as a:

a monthly rate
b weekly rate
c 6-monthly rate
d daily rate
e fortnightly rate
f quarterly rate
(3-monthly rate).

> Divide by 52 for the weekly rate, divide by 2 for the 6-monthly rate and divide by 4 for the quarterly rate.

5 Ava borrowed $4600 from a finance company for 3 years at 18% p.a. interest to buy some furniture for her unit.

a How much interest did she have to pay?
b How much, including interest, did Ava have to repay the finance company?
c What percentage of the amount that Ava repaid the finance company was interest?

6 Martin owed $825 on his credit card. The credit card company charged him one month's interest at 22% p.a.

a How much interest was he charged?
b Calculate the total amount he had to repay the credit card company.

7 Sam borrowed $125 000 from a finance company to set up an online business.
He borrowed the money for 9 months and he was charged 18% p.a. interest.
How much did he have to repay the finance company, including interest?

8 Use the simple interest formula and an equation to solve each problem.

a For how many years will you need to invest $2000 to earn $400 in simple interest at 5% p.a.?
b Natalie earned $1500 interest when she invested $6250 for 3 years.
What was the rate of simple interest p.a.?
c How much do you need to invest to earn $600 in simple interest in 4 years at 5% p.a.?

Example 7

9 When Kyle invested $4600 in simple interest for 3 years his investment grew to $6000.

a How much interest did Kyle's investment earn?
b Calculate the annual rate of simple interest. Answer as a percentage correct to one decimal place.

10 What rate of simple interest is required for $6000 to grow to $9000 in 5 years?
Answer as a percentage correct to one decimal place.

1.03 Compound interest

There's a well-known claim often quoted by financial investors:

Compound interest is the eighth wonder of the world.
Those who understand it, earn it. Those who don't, pay it.

With **compound interest**, you receive 'interest on your interest'. After interest is added to the investment, the next time interest is calculated, it will be based on this larger amount. The amount of interest grows or 'is **compounded**' as the size of the investment grows.

There is also the 'rule of 72'. This rule states that the number of years it takes an investment to double is approximately 72 divided by the percentage rate of compound interest p.a.

EXAMPLE 8

Sarina invested $2000 at 5% p.a. interest compounded annually.

a Calculate the value of Sarina's investment at the end of 3 years.

b Calculate the total interest she earned.

c Approximately how long will it take Sarina's investment to double in size?

Solution

a We need to calculate the interest for each year and add it to the principal before we calculate the following year's interest.

$I = Prn$ where $r = 5\% = 0.05$, $n = 1$ and P changes each year, but $P = 2000$ in the 1st year.

	Interest	Balance
End of the 1st year	$I = \$2000 \times 0.05 \times 1$ $= \$100$	$\$2000 + \$100 = \$2100$
End of the 2nd year	$I = \$2100 \times 0.05 \times 1$ $= \$105$	$\$2100 + \$105 = \$2205$
End of the 3rd year	$I = \$2205 \times 0.05 \times 1$ $= \$110.25$	$\$2205 + \$110.25 = \$2315.25$

At the end of 3 years, Sarina's investment is worth $2315.25.

> Alternatively, calculate the value of 5% interest added to the principal by multiplying by 1.05.
> For example, $2000 × 1.05 = $2100, immediately giving you the first answer in the Balance column.

---

b Total interest
= final balance – original principal.

Total interest = \$2315.25 – \$2000
= \$315.25

OR add the 3 interest amounts in the middle column of the table above.

Total interest = \$100 + \$105 + \$110.25
= \$315.25

Write the answer.

The total interest earned over the 3 years was \$315.25.

c Rule of 72: Doubling time
= 72 ÷ annual interest rate

Doubling time = 72 ÷ 5
= 14.4 years

Write the answer.

Sarina's investment will double in approximately 14.4 years.

Exercise 1.03 Compound interest

1 Jin invested \$10 000 at 5% p.a. interest compounded annually for 3 years. Copy and complete this table to calculate the value of his investment at the end of 3 years.

Example **8**

	Interest	Balance
End of the 1st year	$I = Prn$ $= \$10\,000 \times 0.05 \times 1$ $= \$\_\_\_\_$	\$10 000 + \$____ = \$____
End of the 2nd year	$I = Prn$ $= \$\_\_\_\_ \times 0.\_\_\_ \times 1$ $= \$\_\_\_\_$	\$____ + \$____ = \$____
End of the 3rd year	$I = Prn$ $= \$\_\_\_\_ \times 0.\_\_\_ \times 1$ $= \$\_\_\_\_$	\$____ + \$____ = \$____

2 Zeinab invested \$9000 for 3 years at 5% p.a. compounding yearly.

a How much interest did she earn in the first year?

b How much was in her account at the end of the first year?

c How much was in her account at the end of the second year?

d Calculate the value of her investment at the end of 3 years.

e How much interest will Zeinab earn during her 3-year investment?

f Approximately how long will it take Zeinab's investment to double?

3 Letitia invested $7000 at 4.2% p.a. for 2 years compounding annually.

    a Calculate the total value of her investment at the end of the 2 years.

    b How much interest will she earn during the 2 years?

    c How much less interest would Letitia earn had it been simple interest rather than compound interest?

4 Simon has saved $12 000 from his after-school job that he plans to spend on a car. He is going to invest the money for 3 months until he has his P-plates. Simon's investment is going to pay 0.6% per month, interest compounded monthly.

    a Copy and complete the table.

	Interest	Balance
End of the 1st month	$I = Prn$ $= \$12\ 000 \times 0.006 \times 1$ $= \$\_\_\_\_\_$	$\$12\ 000 + \$\_\_\_\_ = \$\_\_\_\_$
End of the 2nd month	$I = Prn$ $= \$\_\_\_\_ \times 0.\_\_\_ \times 1$ $= \$\_\_\_$	$\$\_\_\_\_ + \$\_\_\_\_ = \$\_\_\_\_$
End of the 3rd month	$I = Prn$ $= \$\_\_\_\_ \times 0.\_\_\_ \times 1$ $= \$\_\_\_\_$	$\$\_\_\_\_ + \$\_\_\_\_ = \$\_\_\_\_$

    b How much interest will Simon make during the 3-month investment?

    c Use repeated multiplication by 1.006 to check your answers to parts **a** and **b**.

5 Keira invested some money in monthly compounding interest. The table shows her interest and balance calculations.

	Interest	Balance
End of the 1st month	$I = Prn$ $= \$9600 \times 0.008 \times 1$ $= \$76.80$	$\$9600 + \$76.80 = \$9676.80$
End of the 2nd month	$I = Prn$ $= \$9676.80 \times 0.008 \times 1$ $= \$77.41$	$\$9676.80 + \$77.41 = \$9754.21$

    a How much money did Keira invest?

    b What was the monthly rate of compound interest?

    c Convert the monthly rate to an annual rate of compound interest.

    d How much interest did Keira's investment earn?

ISBN 9780170413534

6 Last night, Zac received a phone call from an investment advisor he doesn't know telling him about an investment opportunity that is virtually risk-free. The offer is likely to be fully subscribed quickly, so he has to agree to invest immediately. The minimum investment is $10 000 and the interest rate is 15% p.a. compounding annually for 3 years.

   a Calculate the amount of interest the advisor claims $10 000 will receive in the next 3 years.

   b What signs show that this opportunity is most probably a scam?

7 Rachel created a spreadsheet for calculating compound interest that allows her to change the principal and interest rate. This spreadsheet 'Compound interest' is available on NelsonNet for your teacher to download.

Compound interest

	A	B	C	D
1	**Compound interest**			
2	Only enter data in cells shaded blue.			
3				
4	Principal	$1,000.00	Annual rate of interest as a percentage	12%
5				
6				
7		**Account balance at the beginning of the year**	**Interest earned during the year**	**Account balance at the end of the year**
8	Year 1	$1,000.00	$120.00	$1,120.00
9	Year 2	$1,120.00		
10	Year 3			
11	Year 4			
12	Year 5			
13	Year 6			
14	Year 7			
15	Year 8			

   a What formulas could Rachel enter in cells B8, D8 and B9?

   b Rachel used the formula =B8*D4 in cell C8. Why did she use the $ signs?

   c 'Fill down' to complete the spreadsheet calculations for 8 years.

   d Use the spreadsheet to determine the amount of interest a $4500 investment will earn in 8 years at 4.8% p.a. annually compounding interest.

IS IT A SCAM?

Technology and the Internet have changed the way we do things and have made life easier for us. However, technology has also made life easier for criminals to use scams and financial fraud to rob us of our money.

Scamwatch

1 Visit the **Scamwatch** website.

2 Investigate **Unexpected money** scams. List 4 different reasons people fall for scams.

3 Prepare a list of answers you could give to possible scammers when they are trying to convince you to do something you don't want to do.

4 Investigate **Get help**. Describe what you should do if you have sent money to someone you think is a scammer.

1.04 The compound interest formula

It's quicker and easier to calculate compound interest with a formula.

Compound interest

Simple and compound interest

Compound interest

The compound interest formula

$$FV = PV(1 + r)^n$$

where FV = future value or final amount

PV = present value or principal or initial value

r = rate of interest per **compounding period**, expressed as a decimal

n = number of compounding periods.

This is also called the **future value formula**.

This formula can also be written as $A = P(1 + r)^n$.

The interest earned, I, is given by the formula:

$$I = FV - PV$$

EXAMPLE 9

a What will be the final value of an investment of $850 after 3 years at 6.5% p.a. compounding annually?

b How much interest will be earned during the 3 years?

Solution

a Use the formula $FV = PV(1 + r)^n$
$PV = 850, r = 0.065, n = 3$

> Notice in the calculation that $850 is increased by 6.5% 3 times.

$FV = 850 \times (1 + 0.065)^3$
$= 850 \times 1.065^3$
$= 1026.7571 \ldots$
≈ 1026.76

Write the answer.

The final value will be $1026.76

b Interest = final value (FV) – original principal (PV)

Interest = $1026.76 – $850
$= $176.76

Write the answer.

The interest earned during the 3 years will be $176.76.

EXAMPLE 10

Carl invested $3500 for 2 years at 9% p.a. monthly compounding interest to help him save the deposit for a tradesman's truck. How much interest did his investment earn?

Solution

Interest is compounded monthly, so r and n must be in months. Divide 9% = 0.09 by 12.	$r = 0.09 \div 12 = 0.0075$ per month $n = 2 \times 12 = 24$ months
Use the formula $FV = PV(1 + r)^n$ $PV = 3500, r = 0.0075, n = 24$	$FV = 3500 \times (1 + 0.0075)^{24}$ $= 3500 \times 1.0075^{24}$ $= 4187.4473 \dots$ ≈ 4187.45
Interest $= FV - PV$	Interest $= \$4187.45 - \$3500$ $= \$687.45$
Write the answer.	The investment earned $687.45 in interest.

EXAMPLE 11

What present value (principal) must be invested at 6% p.a. for 5 years so that it grows to $12 000?

Solution

Use the formula $FV = PV(1 + r)^n$ to work out PV with $FV = 12\,000, r = 0.06, n = 5$.	$12\,000 = PV \times (1 + 0.06)^5$ $12\,000 = PV(1.06)^5$
Divide both sides of the equation by $(1.06)^5$.	$12\,000 \div (1.06)^5 = PV$ $PV = \dfrac{12\,000}{(1.06)^5}$
Round the present value *up* to the nearest cent.	$PV = \$8967.0980 \dots$ $\approx \$8967.10$
Write the answer.	A present value of $8967.10 is needed.

EXAMPLE 12

Joseph invested \$16 000 in annually compounding interest. 6 years later his investment was worth \$21 440. What annual rate of compound interest did Joseph's investment earn, correct to the nearest percentage?

Solution

Use the formula $FV = PV(1 + r)^n$ to work out r with $FV = 21\ 440$, $PV = 16\ 000$, $n = 6$.

$$21\ 440 = 16\ 000 \times (1 + r)^6$$

Divide both sides of the equation by 16 000.

$$\frac{21\ 440}{16\ 000} = \frac{16\ 000(1+r)^6}{16\ 000}$$

$$1.34 = (1 + r)^6$$

Take the 6th root of both sides.

$$1 + r = \sqrt[6]{1.34}$$

On a calculator, enter 6 [SHIFT] [xʸ] 1.34 [=] (or [x■] instead of [xʸ])

$$1 + r = 1.04998\ldots$$
$$r = 1.04998\ldots - 1$$
$$= 0.04998\ldots$$

Multiply by 100% to convert the interest rate, r, to a percentage.

$$= 0.04998\ldots \times 100\%$$
$$= 4.998\ldots\%$$
$$\approx 5.0\%$$

Write the answer.

Joseph's investment earned 5% p.a. compound interest

Exercise 1.04 The compound interest formula

1 Use the compound interest formula $FV = PV(1 + r)^n$ with $PV = 7600$, $r = 0.075$ and $n = 5$ to calculate the value of FV.

2 David invested \$1870 at 6.8% p.a. compounding annually for 4 years.

  a Write the values for PV, r and n for the compound interest formula.

  b Calculate the value of David's investment at the end of the 4 years.

  c How much interest did David earn?

Example 9

3 Megan won \$5260 in lotto and decided to invest it at 7.1% p.a. compounding yearly for 3 years. How much will her investment be worth at the end of the 3 years?

4 Danielle's bank pays 6.36% p.a. compounding monthly on term deposits.

  a Explain why 6.36% p.a. is equivalent to 0.53% monthly.

  b Danielle invested \$1950 in a term deposit for 18 months. Calculate the value of her investment at the end of 18 months.

  c How much interest will Danielle earn from this investment?

Example 10

5 When Simon retired, he invested $145 600 from his retirement funds at 5.88% p.a. monthly compound interest for 5 years.

   a What monthly interest rate is equivalent to 5.88% p.a.?

   b For how many months did Simon invest the money?

   c How much interest did Simon make from this investment?

6 Calculate the future value of each investment.

> Remember! Interest rate and number of periods must be in the same units as the compounding period.

   a $14 000 invested at 6% p.a. monthly compound interest for 4 years.

   b $1350 invested at 5.4% p.a. compounding monthly for $1\frac{1}{2}$ years.

   c $25 000 invested for 5 years at 7.2% p.a. compounding every 6 months.

   d $60 000 invested for 7 years at 7.2% p.a. compounding quarterly; that is, every three months.

   e $15 000 invested at 6% p.a. for 2 years, interest compounding quarterly.

Example 11

7 Dinesh wants to invest some money so that it will grow to $24 000 in 5 years time, when he will travel through Europe. If the interest rate is 4% p.a., what amount should Dinesh invest, rounded up to the nearest cent?

8 Kristina's investment grew to $6237.15 after 8 years. What was the initial amount of her investment if the interest rate was 8.7% p.a.?

9 Calculate the present value of an investment if $10 000 is required in 6 years' time at 3% p.a. compounded quarterly.

Example 12

10 Use a calculator's [SHIFT] [x^y] keys to solve each equation, and write the value of r as a percentage, correct to 2 decimal places.

   a $(1 + r)^3 = 1.25$ **b** $(1 + r)^7 = 1.96$

   c $2500(1 + r)^8 = 4000$ **d** $6450(1 + r)^6 = 10\ 000$

11 Dion invested $8000, with a future value of $10 000 in 3 years. What is the annual percentage rate of compound interest calculated on the investment, correct to 2 decimal places?

12 Paul heard about an investment opportunity where a $1000 investment grows to $1500 in 2 years.

   a Calculate, correct to one decimal place, the compound interest rate for this investment.

   b Do you think this is a safe investment? Justify your answer.

13 Angie invested $5600 for 3 years and received $950 interest.

   a What was the final value of Angie's investment, including interest, when it matured?

   b Explain why the solution to the equation $6550 = 5600 \times (1 + r)^{36}$ gives the monthly rate of compound interest that Angie's investment earned.

   c Calculate the monthly rate of compound interest as a percentage correct to 2 decimal places.

14 Connor invests $10 000 for 3 years. Which rate will give him the biggest return?

   A 6% p.a. simple interest

   B 5.9% p.a. compounded annually

   C 5.85% p.a. compounded half-yearly

15 Ray has $50 000 to invest for 4 years. Which investment option will give him the better return: 7% p.a. simple interest or 6.2% p.a. compounded monthly?

16 Jill used this spreadsheet to calculate the amount of interest she would earn when she invested $60 000 at 8% p.a. for 4 years with monthly compounding interest.

	A	B	C	D	E	F	G	H
1	Compound interest							
2								
3								
4	Enter the present value in G4						$60,000.00	
5	Enter the number of compounding periods in a year in G5						12	
6	Enter the annual interest rate as a decimal in G6						0.08	
7	Enter the term of the investment in years in G7						4	
8								
9			Final value of the investment =		$82,539.97			
10	Interest earned during the investment =				$22,539.97			
11								

   a One of the formulas Jill used is =(E9-G4). What does this formula calculate?

   b Jill made an error when she entered the formula =G4*(1+G6/G5)^G7*G5 in cell E9. Correct the error in Jill's formula.

   c Construct a spreadsheet similar to Jill's spreadsheet and use your spreadsheet to determine the change in the interest Jill will earn if instead of monthly compounding her investment compounds weekly.

1.05 Compound interest calculators

MoneySmart

We can use technology to calculate compound interest, such as:

- an online compound interest calculator like the one on **The Calculator Site** or **MoneySmart**.

Different compounding periods

- the 'Different compounding periods' spreadsheet on NelsonNet.
- the TVM mode on a graphics calculator.

Compounding periods: spreadsheet

Compound interest table

EXAMPLE 13

Jana invested $2375 for 3 years in monthly compounding interest at 4.5% p.a. Calculate the final value of her investment and the amount of interest she earned.

Solution

Interest calculator

Online compound interest calculator:

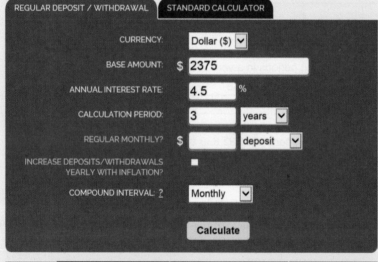

Year	Year Interest	Total Interest	Balance
1	$109.11	$109.11	$2,484.11
2	$114.12	$223.23	$2,598.23
3	$119.36	$342.59	$2,717.59

The Calculator Site.com (thecalculatorsite.com)

Final value of the investment = $2717.59

Total interest = $342.59

Spreadsheet on NelsonNet:

Enter the 4 given values in column B:

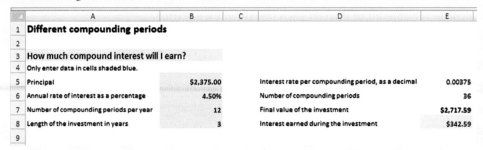

	A	B	C	D	E
1	**Different compounding periods**				
2					
3	**How much compound interest will I earn?**				
4	Only enter data in cells shaded blue.				
5	Principal	$2,375.00		Interest rate per compounding period, as a decimal	0.00375
6	Annual rate of interest as a percentage	4.50%		Number of compounding periods	36
7	Number of compounding periods per year	12		Final value of the investment	$2,717.59
8	Length of the investment in years	3		Interest earned during the investment	$342.59
9					

TVM mode on a calculator:

Using the **TVM mode** on a calculator, enter the values shown.
The value for *FV* is irrelevant.

The final value of the investment is $2717.59.

Interest earned = $FV - PV$

$$= \$2717.59 - \$2375.00$$

$$= \$342.59$$

Exercise 1.05 Compound interest calculators

Use the technology of your choice to answer the questions in this exercise.

Example 13

1 Calculate the amount of interest each investment will earn.

	Principal	Interest rate	Term	Compounding
a	$10 000	6% p.a.	3 years	Monthly
b	$2500	4.2% p.a.	4.5 years	Quarterly
c	$1840	3.6% p.a.	5 years	Half-yearly
d	$3800	4.6% p.a.	2 years	Daily

2 Marissa won $5800 and decided to invest it at 7.2% p.a. monthly compounding interest for 3 years.

  a How much will Marissa's win grow to in 3 years?

  b How much interest will Marissa's investment earn?

3 Brody saved $1660 for a holiday he will take after he finishes Year 12. He invested it at 5% p.a. monthly compounding interest for 1 year. How much will be in his holiday account at the end of the year?

4 Nicky received a $2500 bonus as a result of increased company profits. She is going to invest it for 1 year at 6% p.a. How much more interest will she earn in monthly compounding interest compared to simple interest?

5 When Kris was made redundant at work she was given a lump-sum payment of $40 000. She invested it for 5 years. How much more interest will she earn from daily compounding interest than from annually compounding interest at 6% p.a.?

6 On his retirement, Tan received a lump sum superannuation payment of $450 000 and he is going to invest it for 2 years. He can invest it at 6.4% p.a. monthly compounding or 6.55% p.a. annually compounding interest. Which investment will give him the better return? Justify your answer.

7 Lucy has $200 000 to invest to provide for her retirement. She considers 2 different investment options.

	Investment term	Fees
Finance company	5% p.a. compounding monthly	$10 per month
Managed funds	6% p.a. daily compounding	6.5% of the interest earned

Which option will give Lucy the higher return in 12 months? Explain your answer.

DIFFERENT COMPOUNDING PERIODS

In this investigation, we are going to determine a relationship between the amount of interest earned and the frequency of the interest compounding period.

What you have to do

1 Use the technology of your choice to copy and complete the missing values in the table.

Invest $10 000 for 4 years at 8% p.a. compounding interest	
Compounding period	**Interest earned**
Annually	
Every 6 months	
Monthly	
Weekly	
Daily	

2 Write a sentence to describe the observations you've made.

3 Check the accuracy of your observations by completing a table for another investment.

Invest $50 000 for 7 years at 12% p.a. compounding interest	
Compounding period	**Interest earned**
Annually	
Every 6 months	
Monthly	
Weekly	
Daily	

Chris is going to invest $2000 for 3 years at 6% p.a. compounding interest. He can select either monthly compounding or fortnightly compounding interest. Which compounding period will give him the better return? Give a reason for your answer.

INVESTIGATION

IS COMPOUND INTEREST ALWAYS BETTER THAN SIMPLE INTEREST?

In this investigation, we are going to make some calculations to help us choose which investment has the better return.

What you have to do

1 Copy the tables below.

2 Use the technology of your choice to complete the tables. To make it easier, make the principal $1000 in every calculation.

3 Complete the class discussion questions after you have finished the calculations.

Which is better: simple or compound interest?

Part A: The interest rates and the terms are the same

Simple interest investment	Annually compounded interest investment	Summary
Term: 4 years Interest rate: 5% p.a.	Term: 4 years Interest rate: 5% p.a.	Simple interest = Compound interest = Which investment is better?
Term: 6 years Interest rate: 3% p.a.	Term: 6 years Interest rate: 3% p.a.	Simple interest = Compound interest = Which investment is better?
Term: 20 years Interest rate: 7.5% p.a.	Term: 20 years Interest rate: 7.5% p.a.	Simple interest = Compound interest = Which investment is better?

Part B: The terms are the same but the compound interest rate is higher than the simple interest rate

Simple interest investment	Annually compounded interest investment	Summary
Term: 4 years Interest rate: 5% p.a.	Term: 4 years Interest rate: 5.1% p.a.	Simple interest = Compound interest = Which investment is better?
Term: 6 years Interest rate: 3.75% p.a.	Term: 6 years Interest rate: 4% p.a.	Simple interest = Compound interest = Which investment is better?
Term: 20 years Interest rate: 7.5% p.a.	Term: 20 years Interest rate: 8.1% p.a.	Simple interest = Compound interest = Which investment is better?

Part C: The terms are the same but the simple interest rate is higher than the compound interest rate

Simple interest investment	Annually compound interest investment	Summary
Term: 4 years Interest rate: 5% p.a.	Term: 4 years Interest rate: 4.8% p.a.	Simple interest = Compound interest = Which investment is better?
Term: 6 years Interest rate: 3% p.a.	Term: 6 years Interest rate: 2.75% p.a.	Simple interest = Compound interest = Which investment is better?
Term: 20 years Interest rate: 12% p.a.	Term: 20 years Interest rate: 7.5% p.a.	Simple interest = Compound interest = Which investment is better?
Term: 3 years Interest rate: 12% p.a.	Term: 3 years Interest rate: 7.5% p.a.	Simple interest = Compound interest = Which investment is better?

- When the interest rates and the terms are the same, does simple or compound interest give the better return?

- When the compound interest rate is higher than the simple interest rate, and the terms are the same, which type of investment produces the better return?

- When the simple interest rate is bigger than the compound interest rate, and the terms are the same, will one or the other type of interest always give the better return?

1.06 Inflation

Prices and inflation

Inflation is an increase in the prices of goods and services. One of the challenges facing governments worldwide is controlling inflation. Inflation decreases the value of money.

We can use the future value formula to make price predictions with inflation, and the rule of 72 to calculate the time required for prices to double.

EXAMPLE 14

Inflation is currently at 3% p.a. At the moment, Caroline's favourite take-away meal costs $22.

a Predict the cost of Caroline's favourite meal in 5 years time if inflation stays the same.

b Predict the length of time it will take for the meal to be priced at $44.

Solution

a We can use the formula $FV = PV(1 + r)^n$
$PV = 22, r = 0.03, n = 5$.

$FV = 22 \times (1 + 0.03)^5$
$= 22 \times (1.03)^5$
$= 25.50402 \ldots$
≈ 25.50

Write the answer.

Caroline's favourite take-away meal will cost $25.50.

b $44 is double $22.
Use the rule of 72 with $r = 3\%$.

Doubling time $= 72 \div 3$
$= 24$

Write the answer.

At 3% p.a. inflation, it will take 24 years for the price to double.

Exercise 1.06 Inflation

1 The price of a new small car is $18 000.

   a Calculate the price of a similar new small car in 4 years time if inflation is 3.5% p.a.

   b At 3.5% p.a. inflation, how long will it take the price of the car to double?

Example
14

2 An antique table is valued at $560 and its value is increasing at 7% p.a. How much will the table be worth in 6 years time?

3 From 1998 to 2016, the average rate of Australian wage increases was 3.38% p.a.

   a In 2006, Joe earned $13 per hour. Ten years later, Joe was still doing exactly the same job. How much did he earn per hour in 2016?

   b In 2014, Maggie earned $23/hour. What was her wage in 2016?

4 In 2016, the Australian pay rate for a plumbing foreman was $38/hour and the rate had been increasing by an average of 3.4% p.a.

   a Explain why the solution to the equation $PV \times (1.034)^{10} = 38$ gives the plumbing foreman's pay rate in 2006.

   b Solve the equation to calculate the plumbing foreman's pay rate in 2006.

   c In 2016, the pay rate for an apprentice plumber was $17.14/hour. How much was an apprentice plumber paid per hour in 2006?

5 A hamburger cost $1 in 1971. How much would a hamburger cost in 2036, if the rate of inflation is 5.09% p.a.?

6 Grandma remembers chocolate frogs costing only 10 cents each.

   a At 5% p.a. inflation, how long does it take prices to double?

   b Complete this table.

Cost of a chocolate frog	10c	20c	40c	80c	$1.60
Number of years					

   c Today, chocolate frogs cost 75c each. Use the table of values above to estimate the year when Grandma remembers chocolate frogs costing 10c each.

7 Hyperinflation is when prices rise by 50% or more per month. The currency in Hungary at the end of World War II was called a pengo, and during hyperinflation, prices rose by 207% each day!

Complete this table to show how the cost of a basket of essential food supplies in wartime Hungary increased over 6 days during hyperinflation.

	Monday	Tuesday	Wednesday	Thursday	Friday	Saturday
Cost of basket of food (pengo)	8					

8 Annual rates of inflation can change dramatically when a country is experiencing economic difficulties as they did in Venezuela from 2013 to 2016.

Year	Rate of inflation
2013	41%
2014	63%
2015	121%
2016	481%

The currency in Venezuela is the bolivar. At the beginning of 2013, dinner for two in Caracas, the capital of Venezuela, cost approximately 1000 bolivars.

a Evaluate the expression $1000 \times (1 + 0.41)$ to calculate the cost of dinner for two at the end of 2013.

b Use a series of calculations to show that, by the end of 2016, the cost of dinner for two was almost 30 000 bolivars.

9 Although most items we buy depreciate as they get older, some 'collectibles' appreciate. If you have any collectibles, keep them in their unopened, original box to keep their value as high as possible. This table shows the 2017 value of some prize collectibles.

	Item	Value
a	Certified American baseball card	$150 000
b	Mint condition *Spiderman* comic book	$20 000
c	Vintage *Star Wars* costume in original packet	$18 000
d	Original Apple computer in unopened box	$670 000
e	Limited edition *Frozen* doll in unopened box	$2500
f	Original copy of one of Da Vinci's notebooks	$50 000

Assuming an annual rate of appreciation of 5%, calculate a predicted 2020 value of each collectible listed in the table. Answer to the nearest dollar.

10 Do an online search for collectibles. Write a list of items that are currently valuable. Which collectibles are increasing in value more quickly than others?

No-one knows the future rates of **appreciation** of collectibles, nor which particular items will become valuable collectibles.

SYDNEY HOUSE PRICES

Despite making accurate calculations, we can never predict prices with 100% accuracy. In this investigation, we are going to make some predictions based on 2017 prices and then check the predictions against what actually happened.

In 2017, the median house price in Sydney was $1.15 million, increasing at 11% p.a.

1 Copy and complete this table of values to show the predicted median Sydney house price.

Year	2017	2018	2019	2020	2021
Median price	$1 150 000				

2 Use the rule of 72 to predict when the median house price will become close to 2 million dollars.

3 In 2017, inflation was only 2% p.a., much lower than the increase in house prices. Copy and complete this table to predict house prices changing as a result of inflation only.

Year	2017	2018	2019	2020	2021
Median price	$1 150 000				

4 Use an online search to find out the actual median Sydney house price from 2017 until now.

5 Construct a graph showing the predicted values in the tables in Questions **1** and **3**. Then, show the true median prices from your online research.

Discussion questions

1 How accurately did the values you calculated in Questions **1** and **3** predict actual median house prices?

2 Did the 11% p.a. 2017 rate of increase continue? Justify your opinion.

3 In 2017, some financial experts were predicting a crash in Sydney house prices. Did a crash happen? If so, when and what affect did it have?

4 List some factors that you think affect house prices. Justify your opinion.

WORD MATCH

Match the terms in the left column with their correct meanings in the right column.

	Term		Meaning
1	interest	A	Prices increase and the value of money decreases
2	compound interest	B	The future value (compound interest) formula
3	compounding period	C	The rate of return on an investment, usually annually and expressed as a percentage
4	inflation	D	This interest is calculated on the original investment only
5	scam	E	How often the interest is calculated using compound interest
6	hyperinflation	F	The original amount of an investment
7	$I = Prn$	G	An illegal trick for cheating you of your money or information
8	interest rate	H	The amount earned on an investment
9	principal	I	The simple interest formula
10	$FV = PV(1 + r)^n$	J	This interest is calculated on the current value of the investment, including interest previously added to the principal
11	simple interest	K	When prices rise by 50% or more in a month

SOLUTION TO THE CHAPTER PROBLEM

Problem

Ryan invested $25 000 for 3 years, earning compound interest at 4% per year.

Melanie also invested $25 000 for 3 years, earning simple interest at 4.2% per year.

Whose investment earns more interest?

Solution

Ryan's investment

$FV = PV(1 + r)^n$

$PV = 25\ 000$, $r = 0.04$ and $n = 3$

$$
\begin{aligned}
\text{Final amount} &= 25\ 000 \times (1 + 0.04)^3 \\
&= 25\ 000 \times (1.04)^3 \\
&= 28\ 121.60
\end{aligned}
$$

Ryan's investment will be worth $28 121.60 at the end of 3 years.

$$
\begin{aligned}
\text{Ryan's interest} &= \$28\ 121.60 - \$25\ 000 \\
&= \$3121.60.
\end{aligned}
$$

Melanie's investment

Simple interest $= Prn$

$P = 25\ 000$, $r = 0.042$ and $n = 3$

$$
\begin{aligned}
\text{Melanie's interest} &= 25\ 000 \times 0.042 \times 3 \\
&= \$3150.
\end{aligned}
$$

Melanie's investment will earn $3150 interest.

Melanie's investment will earn more interest ($28.40 more than Ryan's).

- What part of this chapter will be most useful to you in the future?
- Did any of the information in this chapter surprise you? What part?
- How are you going to avoid falling for a financial scam?
- Are there any parts of this chapter that you don't understand? If yes, ask your teacher for help.

Copy and complete this mind map of the topic, adding detail to its branches and using pictures, symbols and colour where needed. Ask your teacher to check your work.

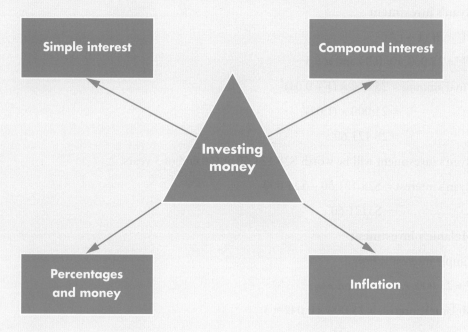

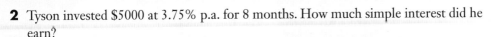

1. TEST YOURSELF

1 Jacinta paid $62 for some gym equipment she bought online, then she sold it for $98.

   a How much profit did she make?

   b Calculate her percentage profit. Answer to the nearest whole percentage.

Exercise 1.01

2 Tyson invested $5000 at 3.75% p.a. for 8 months. How much simple interest did he earn?

Exercise 1.02

3 When Con invested $8200 for 3 years, he earned $984 in simple interest. Calculate the annual rate of interest Con earned on his investment.

Exercise 1.02

4 Ellen invested $4000 at 5% p.a. interest compounded annually for 3 years. Copy and complete this table to calculate the value of her investment at the end of 3 years.

Exercise 1.03

	Interest	Balance
End of the 1st year	$I = Prn$ $= \$4000 \times 0.05 \times 1$ $= \$_____$	$\$4000 + \$\_\_\_\_ = \$_____$
End of the 2nd year	$I = Prn$ $= \$_____ \times 0.\_\_\_ \times 1$ $= \$\_\_\_\_$	$\$\_\_\_\_ + \$\_\_\_\_ = \$\_\_\_\_$
End of the 3rd year	$I = Prn$ $= \$\_\_\_\_ \times 0.\_\_\_\_ \times 1$ $= \$\_\_\_\_$	$\$\_\_\_\_ + \$\_\_\_\_\_ = \$_____$

5 a Use the compound interest formula to calculate the future value of a $6400 investment annually compounding at 5% p.a. for 4 years.

Exercise 1.04

   b How much interest will the investment earn?

6 Shelly invested $3000 at 4.8% p.a. monthly compounding interest for 3 years.

Exercise 1.04

   a What is the monthly interest rate?

   b How much interest will Shelley earn?

7 Pedro has $10 000 to invest for 6 years. Use the technology of your choice to determine the best investment: 4% p.a. monthly compounding, 4.1% p.a. annually compounding or 4.4% simple interest?

Exercise 1.05

8 A tablet device is currently valued at $340. Predict the value of the device in 5 years time assuming an average 3% p.a. inflation rate.

Exercise 1.06

9 Inflation is currently at an average rate of 3.2% p.a. At present, Harry earns $25/hour. How much was his hourly wage 6 years ago?

Exercise 1.06

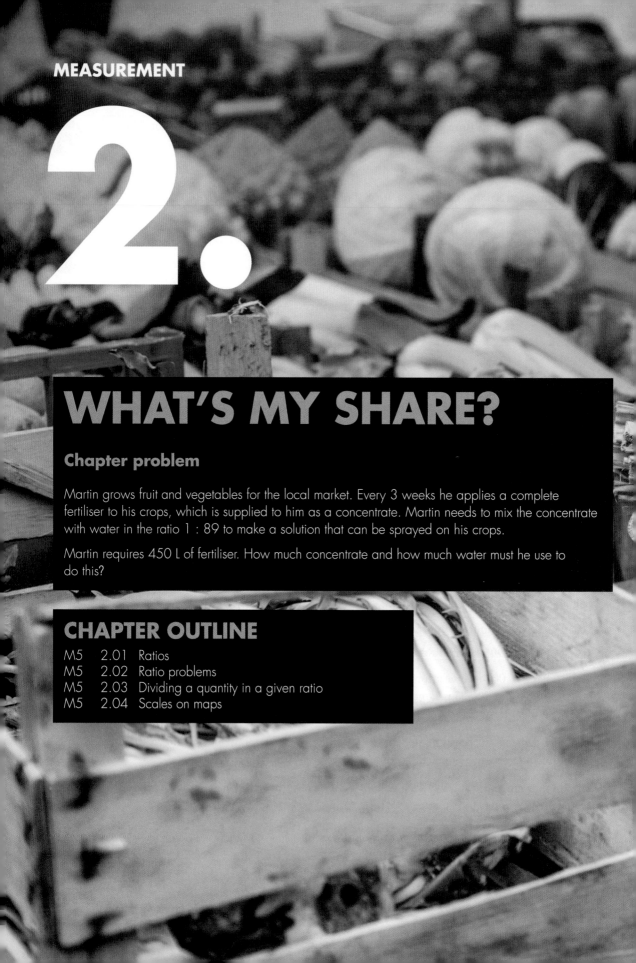

MEASUREMENT

2.

WHAT'S MY SHARE?

Chapter problem

Martin grows fruit and vegetables for the local market. Every 3 weeks he applies a complete fertiliser to his crops, which is supplied to him as a concentrate. Martin needs to mix the concentrate with water in the ratio 1 : 89 to make a solution that can be sprayed on his crops.

Martin requires 450 L of fertiliser. How much concentrate and how much water must he use to do this?

CHAPTER OUTLINE

WHAT WILL WE DO IN THIS CHAPTER?

- Write ratios and simplify them
- Solve problems involving ratios
- Divide quantities in a given ratio
- Solve problems involving scales on maps

HOW ARE WE EVER GOING TO USE THIS?

- When measuring while cooking, building and renovating
- When mixing fertiliser, hair colour or other concentrates
- When sharing the rent payment or profit made between people in a fair way
- When reading maps and making travel plans

Ratios

Ratio word
blanks

Ratio
match-up
puzzle

2.01 Ratios

A **ratio** shows the relative sizes of two or more quantities of the same type. Before we can write quantities as a ratio they must be in the **same units**.

EXAMPLE 1

Write the ratio for:

a   15 cm to 37 cm   **b**   59 seconds to 2 minutes.

Solution

a   Both quantities have the same units, cm.   Ratio is 15 : 37.

b   These quantities have different units, so   2 minutes = 2 × 60 = 120 seconds.
convert 2 minutes to seconds.

      Now, both quantities are in seconds.   Ratio is 59 : 120.

To **simplify** a ratio, divide each term in the ratio by the same amount.

EXAMPLE 2

Simplify each ratio.

a   10 : 6   **b**   21 : 7   **c**   10 : 15 : 20

Solution

a   Both 10 and 6 can be divided by 2. $10 : 6 = 10 \div 2 : 6 \div 2$
$$= 5 : 3$$

> We can also use the fraction key on our calculator: enter 10 $\boxed{a^b/_c}$ 6 $\boxed{=}$ and the calculator will display $\frac{5}{3}$, which we write as 5 : 3.

b   Divide both 21 and 7 by 7. $21 : 7 = 21 \div 7 : 7 \div 7$
$$= 3 : 1$$

> For 21 $\boxed{a^b/_c}$ 7 $\boxed{=}$, the calculator will display 3, which we can write as 3 : 1.

c   All 3 terms can be divided by 5. $10 : 15 : 20 = 10 \div 5 : 15 \div 5 : 20 \div 5$
$$= 2 : 3 : 4$$

> When there are more than two terms in the ratio, we can't use the calculator to simplify.

> We should always check our answer to see if it can be simplified further.

   NCM 12. Mathematics Standard 1   ISBN 9780170413534

Exercise 2.01 Ratios

1 Write the ratio for:

Example
1a

   a   13 g to 25 g   **b** $17 to $100   **c**   5 litres to 12 litres

   d   150 km to 1 km   **e**   31 seconds to 51 seconds to 101 seconds.

2 Marianne surveyed the eye colour of the 25 members of her netball club and found that 12 had brown eyes, 8 had blue eyes, 4 had green eyes and one had hazel eyes. Write the ratio of players with:

   a   hazel eyes to green eyes   **b**   brown eyes to hazel eyes

   c   blue eyes to brown eyes   **d**   brown eyes : green eyes

   e   blue eyes : brown eyes : hazel eyes   **f**   green eyes to blue eyes to brown eyes.

3 Kris buys avocados at the market for $2 each and sells them in his grocery store for $3 each.

   a   How much profit does Kris make on each avocado?

   b   What is the ratio of:

      i   selling price to cost price?   **ii**   selling price to profit?

      iii   cost price to profit?   **iv**   profit to selling price?

4 Of the 75 people who attended a bush dance, 27 were men, 29 were women and the remainder were children. Find the following ratios:

   a   men to women   **b**   children to men   **c**   women to the total

   d   women to children   **e**   men to women to children.

Alamy Stock Photo/Stephen Saks

5 Write the ratio for:

a 3 cents to $2.50

b 23 mm to 3 cm

c 37 minutes to 2 hours

d 10 years to 36 months

e 14 days to 5 weeks

f 0.3 km to 173 metres.

6 The diagram shows a cluster of yellow, green, purple, grey, red and black balls.

Write the colours that are in the ratio:

a 6 : 1

b 4 : 3

c 2 : 5

d 2 : 21

e 2 : 1 : 6

f 3 : 4 : 5

> Be sure to write the colours in the same order as the ratio is written.

7 Simplify each ratio.

a 8 : 6

b 9 : 12

c 15 : 10

d 20 : 10

e 21 : 14

f 27 : 9

g 18 : 12

h 150 : 25

i 120 : 24

j 56 : 49

k 28 : 52

l 200 : 50

m 8 : 12 : 4

n 6 : 3 : 18

o 15 : 25 : 50

p 90 : 60 : 120

8 Simplify each ratio using your calculator.

a 4.5 : 9

b 1.8 : 2.7

c $2\frac{1}{2} : 7\frac{1}{2}$

d $14 : 3\frac{1}{2}$

e $\frac{1}{2} : \frac{3}{4}$

f 2.9 : 5.8

g 3.5 : 2.1

h 3.6 : 1.2

9 Karen uses a soil conditioning solution on her vegetable garden. The ingredients are listed on this label shown.

Express in simplest form the ratio of:

a seaweed extract to water

b organic acid to water

c seaweed extract to organic acid

d seaweed extract to total solution volume

e seaweed extract to organic acid to water.

FERTILISER

Ingredients
Seaweed Extract 135 mL
Organic Acid   45 mL
Water   900 mL

10 The table shows the average annual yield in kilograms from various fruit and nut trees.

Standard apple	144		Nectarine	63
Apricot	36		Fig	9
Standard pear	54		Walnut	27

What is the yield ratio of:

a apple to pear?

b nectarines to walnuts?

c apricot to apple?

d pear to fig?

e apples to nectarines to figs?

11 Steven runs a small gourmet business producing specialty fruits, vegetables and cereal products. He also has poultry producing a variety of eggs. The drawing shows the area of land he has dedicated to each activity. Find the ratio of the following production areas.

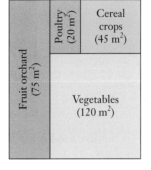

a fruit to vegetables

b egg to total area

c vegetables to total crop area (the area excluding poultry)

d cereal crop to the total crop area

e vegetable to fruit to cereal crop area.

12 Robyn is cleaning the floor with a cleaning liquid she has mixed from a detergent concentrate and water. The instructions say to mix concentrate to water in the ratio 3 : 16 for light cleaning or 5 : 16 for heavy-duty cleaning. Robyn has mixed 22.5 mL of concentrate with 72 mL of water for this job.
Is she doing light or heavy-duty cleaning?

2.02 Ratio problems

When we prepare recipes, an infant's food formula, a hair dye or a fertiliser for the garden, it is important that the ingredients are mixed in the required ratio. Too much or too little of one ingredient can cause damage or have unintended effects. Most ratio problems can be solved using the **unitary method**.

Shutterstock.com/monticello

A poultry food is made by combining grain, protein meal and vegetable matter in the ratio 5 : 2 : 1.

a Alex has 20 kg of grain to make poultry food. How much protein meal and vegetable matter will he need?

b How much poultry food will this make?

Solution

a	5 : 2 : 1 means 5 parts grain, 2 parts protein meal and 1 part vegetable matter. 5 parts of grain is 20 kg. Divide 20 by 5 to find 1 part.	5 parts = 20 kg 1 part = 20 ÷ 5 = 4 kg
	Multiply by 2 to find 2 parts protein meal.	Protein meal = 2 × 4 = 8 kg
	The vegetable matter is 1 part only.	Vegetable matter = 4 kg
		8 kg of protein meal and 4 kg of vegetable matter need to be mixed with 20 kg of grain.
b	Add the ingredients.	Total weight of poultry food = 20 + 8 + 4 = 32 kg

   ISBN 9780170413534

Exercise 2.02 Ratio problems

1 To make a fertiliser for his rose garden, Ahmed mixes manure and grass clippings in the ratio 3 : 2.

a How many spadefuls of grass clippings does Ahmed need to mix with 15 spadefuls of manure?

b If he uses 6 spadefuls of grass clippings, how much manure will he need to mix with it?

2 Nickel brass is a metal alloy containing copper, zinc and nickel mixed in the ratio 14 : 5 : 1. Nickel brass is used in many musical instruments and is the metal used to make the outer section of the British pound coin.

Shutterstock.com/Jeanette Teare

a A metal worker used 70 g of copper to make a batch of nickel brass. How much zinc and nickel did he require?

b Van has 8 g of nickel. How much copper and zinc will he need to mix with it to make nickel brass?

c How much nickel brass can Van make?

3 When Padmina makes scones for the school fete, she mixes self-raising flour and milk in the ratio 3 : 1 by volume.

a How much milk would she add to 15 cups of flour?

b If she uses 6 cups of milk, how many cups of flour are required?

4 Felix makes saddles and other leather goods. He treats them with a leather dressing made up from lanolin and neatsfoot oil mixed in the ratio 2 : 3.

a How much neatsfoot oil should Felix mix with 40 mL of lanolin to make his leather dressing?

b How much leather dressing will this make?

c How much leather dressing is Felix making when he uses 45 mL of neatsfoot oil?

5 Abbie mixes a spray to kill weeds by mixing concentrated weed killer with water.

Weed to be controlled	Ratio of concentrate to water
Flat leaf weeds	1 : 10
General perennial weeds	3 : 20
Woody weeds	1 : 50

a Abbie needs to kill some blackberries, which are woody weeds. If she uses 40 mL of the concentrate, how much water should she add to make the spray?

b How many litres of water is this?

c Abbie is going to clear a small area to make a new garden. It is presently infested with perennial weeds. How much concentrate does she need to add to 1 L of water to make a spray to kill the weeds?

> First, convert 1 L to mL.

d Abbie is going to make a spray to use on the flat-leaved weeds in her lawn. How much concentrate should she mix with 250 mL of water to make the spray?

e Abbie made a spray using 45 mL of concentrate mixed with 300 mL of water. For what type of weed will she use this?

6 Nadim makes garden compost by mixing his vegetable and plant waste with straw and soil in the ratio 12 : 3 : 2. Find the amount of:

a straw he will need if he has 480 L of vegetable and plant waste

b soil that should be added to a mixture containing 12 L of straw

c compost that would be made by mixing 60 L of straw with the other ingredients.

7 Gemma has a vegetable garden and a fruit orchard. To fertilise these, she makes mixtures containing nitrogen (N), phosphorus (P) and potassium (P). The proportions used for the vegetable garden and for the orchard are different. The table shows the ratios Gemma uses.

Plant	N : P : K	Nitrogen (kg)	Phosphorus (kg)	Potassium (kg)	Total mixture (kg)
Vegetables	6 : 4 : 5	54			
				35	
Fruit trees	7 : 1 : 2		15		
		140			

Copy and complete the table to show the weights of different ingredients Gemma uses to make her fertilisers.

ISBN 9780170413534

8 A recipe for making pizza bases mixes plain flour, water and yeast in the ratio
16 : 10 : 1 by weight.

a Jasmine is making pizza dough using 400 g of flour. How many grams of water and yeast will Jasmine need to add?

b Spiro makes all the dough for his pizza shop in the morning. He calculates that he needs to mix 2.5 kg of water with the other ingredients. How much flour and yeast does he use?

c A pizza company uses 3 kg of yeast each day. How much flour and water does the company use?

9 Ryan is renovating his house. He uses a mortar made by mixing cement, sand and lime. Mortars are mixed in different proportions depending on the type of work being done.

Job	Cement : lime : sand in mortar
New brick construction	1 : 0 : 4
Old brick renovation	2 : 3 : 10
Rendering	1 : 1 : 6

a On what type of job does Ryan not use any lime in his mortar?

b Ryan constructed a new brick fireplace. To make the mortar, he used 8 buckets of sand. How many buckets of cement did he need to add?

c Ryan must now render the new fireplace. The mortar he uses contains 12 buckets of sand. How much lime and cement will be required?

d Ryan is going to repair one of the walls of his house using old bricks. How much sand and lime will he need if he mixes his mortar using 4 buckets of cement?

RATIOS OF BODY PARTS

Artists use body part ratios to make sketches of people look realistic. The diagram shows some of the important ratios for sketching images of people.

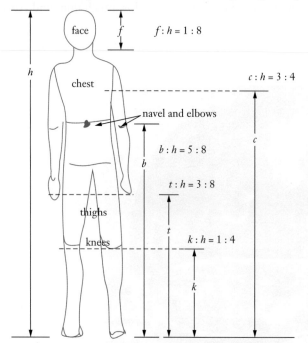

Men and women's bodies have different shapes. Women's necks and waists are thinner whereas their hips and thighs are wider. This table shows the different ratios for adult males and females.

	Males	Females
Shoulder width : head length	7 : 3	2 : 1
Waist width : head length	15 : 7	1 : 1
Hip width : head length	2 : 1	3 : 2

1 Lara is sketching a man. In her sketch, the man is 20 cm tall. How big should she make:

a his head length?

b the height of his chest?

c the width of his shoulders?

2 Dylan is going to make a sketch of a man 12 cm tall. Calculate the size of each body part on the sketch.

   a   length of head   **b**   height of knees

   c   height of mid-thighs and fingertips   **d**   height of navel and elbows

   e   height of chest   **f**   width of shoulders

   g   width of hips

3 In the same sketch, Dylan is going to draw a woman 10 cm tall. Calculate the size of each body part on the sketch.

   a   length of head   **b**   height of knees

   c   height of mid-thighs and fingertips   **d**   height of navel and elbows

   e   height of chest   **f**   width of shoulders

   g   width of hips

4 Use the answers to Questions **2** and **3** to sketch a drawing of the man and woman.

2.03 Dividing a quantity in a given ratio

Sometimes, we want to divide or share quantities in unequal amounts. For example, the owners of businesses will share the profits and the expenses according to the sizes of their shares in the business.

Ratio problems

EXAMPLE 4

Janet and Darryl own a small market garden, which made a profit this year of $135 000. They have agreed to share the profit in the ratio 2 : 3. How much does each person receive?

Calculating quantities in ratios

Solution

The profits are to be divided into 2 parts for Janet and 3 parts for Darryl.

Total number of parts = 2 + 3 = 5.

$$\text{Janet's share} = \frac{2}{5} \times \$135\,000$$
$$= \$54\,000$$
$$\text{Darryl's share} = \frac{3}{5} \times \$135\,000$$
$$= \$81\,000$$

Check the answer. Do the separate amounts sound reasonable and add up to the total?

Checking: $54 000 + $81 000 = $135 000

Write the answer.

Janet receives $54 000 and Darryl receives $81 000.

Ratios in other contexts

ISBN 9780170413534

Hamish needs to mix 180 kg of concrete. Concrete is made using cement, sand and gravel in the ratio 1 : 2 : 3. How much of each product will Hamish need?

Solution

The concrete is made using 1 part cement, 2 parts sand and 3 parts gravel.

Total number of parts = 1 + 2 + 3 = 6.

$$\text{Cement} = \frac{1}{6} \times 180$$
$$= 30 \text{ kg}$$
$$\text{Sand} = \frac{2}{6} \times 180$$
$$= 60 \text{ kg}$$
$$\text{Gravel} = \frac{3}{6} \times 180$$
$$= 90 \text{ kg}$$

Check the answer.

Checking: 30 + 60 + 90 = 180

Write the answer.

Hamish will need 30 kg cement, 60 kg sand and 90 kg gravel.

Exercise 2.03 Dividing a quantity in a given ratio

1 Divide $2100 in the ratio:

   a 2 : 3   **b** 5 : 1   **c** 4 : 2 : 1   **d** 3 : 4 : 7

2 Divide:

   a 855 L in the ratio 1 : 4   **b** 35 lollies in the ratio 4 : 3

   c 24 300 m^2 in the ratio 4 : 5   **d** 819 kg in the ratio 7 : 2

   e 52 min in the ratio 3 : 4 : 6   **f** $2800 in the ratio 8 : 1 : 5

3 Each week, Michael produces 3.5 tonnes of fruit to be sent to the markets. He will send 4 parts to the central Sydney market and 3 parts to a local market. How much will he send to the local market?

4 St Judith's College has 741 students. The ratio of boys to girls is 7 : 6. How many boys are at the school?

5 Jemma is making a bracelet using silver beads, porcelain balls and crystal eyedrops in the ratio 4 : 3 : 2. She uses 45 pieces to make the bracelet. How many of each type will she require?

6 A flat white coffee contains black espresso coffee and milk in the ratio 1 : 2. How much milk is there in a 240 mL mug of flat white coffee?

7 Rose gold (or red gold) is used in jewellery and high-quality flutes because of its attractive appearance and its resonant qualities. 18-carat rose gold is an alloy of gold, copper and silver blended in the ratio 15 : 4 : 1.

   a How much silver would there be in a 40 g piece of rose gold jewellery?

   b A flute made of rose gold weighs 420 g. How much gold and copper would it contain?

Shutterstock.com/Boris Medvede-

8 Sam makes mocktails using 3 parts grape juice, 2 parts cranberry juice and 4 parts sparkling mineral water.

Example **5**

   a How many millilitres of grape juice will he need to make two 225 mL mocktails?

   b How much mineral water will Sam use if he makes 900 mL of the mocktail?

   c How much cranberry juice will this mixture require?

9 Jenny mixes an insecticide concentrate with water before she sprays it on vegetable crops. The mixing ratio depends on the type of vegetable as shown in the table. Jenny is going to make 3 L of each type of spray.

Type of crop	Ratio of concentrate to water	mL of concentrate required	mL of water required
Broccoli	1 : 14		
Cauliflower	1 : 19		
Leafy vegetables	1 : 9		

   a Express 3 L in millilitres.

   b Copy and complete the table to show the number of millilitres of concentrate and water Jenny needs to use for each vegetable.

10 Filomena is a hairdresser who needs to dye a customer's hair. The colour she wants can be made by mixing two dyes A and B in the ratio 5 : 3. Filomena needs to make 120 mL of the mixture. How much of each dye will she need to use?

11 Boris has two machines on his farm that use 2-stroke fuel, a mixture of petrol and oil. The table shows the petrol to oil ratio required for each machine.

Machine	Petrol : oil
Ride-on mower	24 : 1
Cultivator	39 : 1

  a Boris mixes the fuel for the ride-on mower in a 15 L drum. How much petrol must he use?

  b How much oil is required (in litres)?

  c Express your answer to part **b** in mL.

  d For the cultivator, Boris makes the fuel in batches of 20 L. How much oil does he require for each batch? Give your answer in mL.

12 Tina mixes a fertiliser combining nitrogen, phosphorus and potassium in the ratio 5 : 1 : 2. How much of each ingredient will she use to make 120 kg of the fertiliser?

INVESTIGATION

RATIOS AND TV SCREENS

1 Measure the width and height of a variety of TV screens. Alternatively, find the measurements of TV screens on the Internet.

2 Calculate the width : height ratio for each TV. Comment on any similarities and differences.

3 The distance you sit from the TV and the length of the TV's diagonal should be in a ratio 3 : 1 or larger. What is the length of the diagonal of the largest TV you should have in your lounge room?

ISBN 9780170413534

2.04 Scales on maps

Maps usually have a **scale** drawn or written as a ratio.

EXAMPLE 6

The scale on a bushwalking map is 1 : 5000.

a On the map, a walking trail is 4.5 cm long. How long is the actual trail in metres?

b What is the scaled distance (in cm) on the map for a pond that is 750 m long?

Solution

a The scale 1 : 5000 means that 1 unit on the map is equivalent to 5000 units in real life. Multiply the scaled length on the map by 5000.

Actual length = 4.5 cm × 5000
= 22 500 cm

Convert to metres by dividing by 100.

22 500 cm ÷ 100 = 225 m

Write the answer.

The trail is 225 m long.

b Change the actual distance to centimetres.

Actual distance = 750 m
= 750 × 100
= 75 000 cm

Divide by the scale.

Scaled distance = 75 000 ÷ 5000
= 15 cm

Write the answer.

The trail is 15 cm on the map.

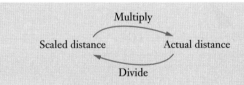

Multiply

Scaled distance   Actual distance

Divide

Exercise 2.04 Scales on maps

Example 6

1 A map's scale is 1 : 2000. Calculate the actual lengths of each feature given their scaled lengths.

   a Lake 1.5 cm wide

   b Walking track 5.8 cm long

   c Creek 9.5 cm long

2 A bushwalking map has a scale of 1 : 25 000. Find:

   a the actual distance for each scaled distance on the map

      i 3 cm   **ii** 5.5 cm   **iii** 9.5 cm

   b the scaled distance on the map for each actual distance

      i 10 km   **ii** 0.5 km   **iii** 4 km

3 The town of Gilgandra is 66 km north of Dubbo. A map uses a scale of 1 : 200 000. How long is the scaled distance between the two towns?

4 Lord Howe Island is 2.8 km wide. A map has a scale of 1 : 50 000. What is the width of Lord Howe Island on the map?

5 This map shows the walking trails that start at the Lodge.

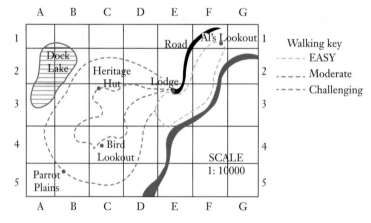

   a What is the scale?

   b What length is represented by 10 cm on the map?

   c How long is the creek?

   d Calculate the length of the moderate walking track.

> You can use a chain or a piece of string to help you find lengths on a map.

   e Jude started walking along the moderate walking track at 11 a.m. He is walking at a speed of 3 km/h. When will Jude finish walking the track?

6 The red star on the map shows the position of Zack's house. Use the scale 1 : 16 000 to calculate the distance Zack walks (along roads) to Tuggerah Lakes Secondary College (shown by the blue star) each day.

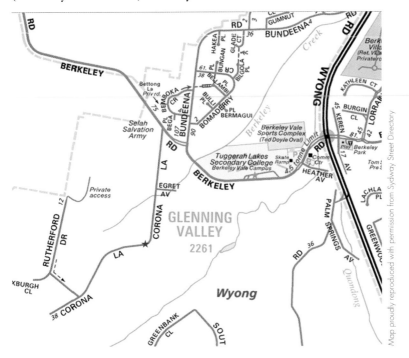

7 This map shows the road around East MacDonnell National Park in Central Australia. The scale is 1 : 1 000 000.

a Use the scale to calculate the length of the road (the circuit outlined in dark blue).

b The suggested time to drive one circuit on this road is $5\frac{1}{2}$ hours. At what average speed do drivers need to travel to complete the circuit in $5\frac{1}{2}$ hours? Express your answer correct to the nearest km/h.

c What do you think the surface of the road is like? Give a reason for your answer.

8 Cadastral maps show the boundaries of properties. This part of a cadastral map shows the dimensions of a block of land and the buildings on the land.

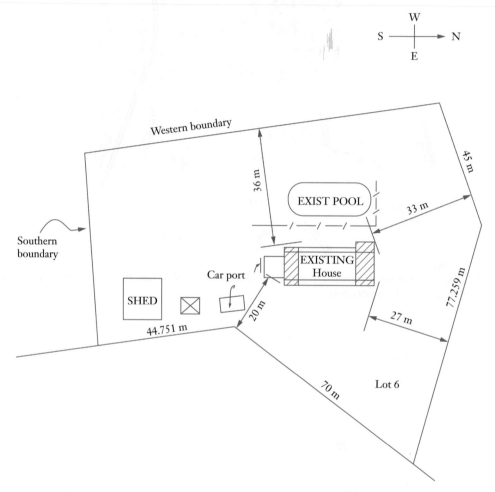

a Approximately how close is the house to the nearest boundary?

b The scale is 1 : 1000. Calculate the length of the southern and western boundaries.

c The owners are going to replace the fence on the southern and western boundaries. The fencing costs $28/m for materials and labour. Calculate the cost of replacing the fencing.

ISBN 9780170413534

9 Mary Street in Gympie, South-East Queensland is prone to flooding. The map shows the heights of floodwaters when the Mary River at Kidd Bridge reaches 18 m and when the Mary River at Kidd Bridge reaches 23.5 m. The scale is 1 : 1000.

—— when the river at Kidd Bridge reaches 18 m   —— 1-in-50 year flood level 23.5m at Kidd Bridge

a Which properties in Mary Street flood when the Mary River reaches 18 m at Kidd Bridge?

b Estimate the height of the Mary River at Kidd Bridge when property 187 Mary St is flooded.

c Calculate the length of Reef St that is under water when the Mary River is at 18 m at Kidd Bridge.

d Calculate the length of Mary Street that is under water during a 1-in-50 year flood.

e Rhys is considering buying the hotel at 135 Mary Street. What recommendation could you give him?

Newspix/Nathan Richter

10 Jose, Isaac and Chase were hiking along The Coast track in the national park. They left their bikes at Otford lookout and set out for Garie Beach.

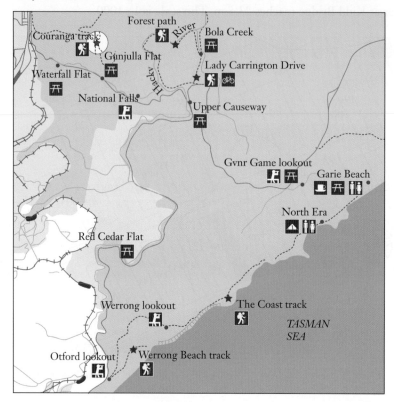

a The map scale is 1 : 80 000. How long is the walk from the Otford lookout to Garie Beach?

b The boys hike at an average speed of 4 km/h. How long should the hike take them?

c When they had been hiking for 2 hours, Chase lost his footing, fell and broke his leg. Isaac phoned emergency services for help. Approximately where on the track should he tell the rescue helicopter they are located?

Peter Rae/Fairfax syndication

RATIO AND MAP SKILLS

In your own words, write a short paragraph explaining how to carry out each of the following skills that you have learned in this chapter.

- Simplifying ratios

- Using the unitary method to solve ratio problems

- Dividing a quantity in a given ratio

- Finding an actual measurement from a map

- On a map, finding a scaled measurement from an actual measurement

SOLUTION TO THE
CHAPTER PROBLEM

Problem

Martin grows fruit and vegetables for the local market. Every 3 weeks he applies a complete fertiliser to his crops, which is supplied to him as a concentrate. Martin needs to mix the concentrate with water in the ratio 1 : 89 to make a solution that can be sprayed on his crops.

Martin requires 450 L of fertiliser. How much concentrate and how much water must he use to do this?

Solution

The 450 L of solution is to be made up of concentrate to water in the ratio 1 : 89.

Total number of parts = 1 + 89 = 90.

$$\text{Concentrate required} = \frac{1}{90} \times 450 \text{ L}$$
$$= 5 \text{ L}$$

$$\text{Water required} = \frac{89}{90} \times 450 \text{ L}$$
$$= 445 \text{ L}$$

Check: 5 + 445 = 450.

2. CHAPTER SUMMARY

- List the parts of this topic you remembered from Years 9–10.
- Give examples of jobs where you would use ratios to mix quantities.
- Write any difficulties you had with the work in this chapter. Ask your teacher to help you.

Copy and complete this mind map of the chapter, adding detail to its branches. Use pictures, symbols and colour where needed. Ask your teacher to check your work.

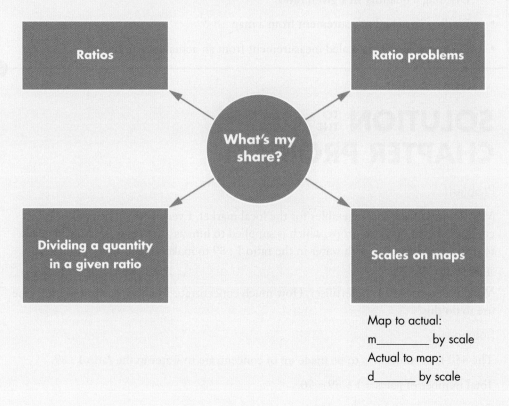

Ratios

Ratio problems

What's my share?

Dividing a quantity in a given ratio

Scales on maps

Map to actual:
m_____ by scale
Actual to map:
d_____ by scale

2. TEST YOURSELF

1 a To paint a picture of a tree, Yvette mixed 5 mL of red paint with 9 mL of green paint to make brown paint. What is the ratio of green paint to red paint in the brown mixture?

Exercise 2.01

b To make a blue-green paint for the leaves, Yvette mixed 6 mL of green paint with 2 mL of blue paint. What is the ratio of green : blue paint in the blue-green mixture? Simplify your answer.

2 To paint the woodwork in his lounge room, Tri added 5 mL of brown paint to 2 L of white paint.

Exercise 2.01

a What is the ratio of brown to white paint in the woodwork mixture?

b Write your answer to part **a** in its simplest form.

3 Simplify each ratio.

Exercise 2.01

    a 4 : 10   **b** 6 : 9

    c 20 : 10   **d** 18 : 12

    e 15 : 10   **f** 12 : 6

    g 100 : 20   **h** 25 : 100

4 Simplify each ratio.

Exercise 2.01

> First, remember to make the units the same.

    a 40 mL to 1 L   **b** 3 L to 20 mL

    c 2 L to 500 mL   **d** 50 cm to 1 m

    e 3 m to 60 cm   **f** 5 kg to 500 g

    g 2 hours to 40 minutes   **h** 45 minutes to half an hour

5 Lara wants to make a lighter shade of orange by mixing yellow and orange paint in the ratio 3 : 1. How much orange will she add to 15 mL of yellow paint?

Exercise 2.02

Exercise
2.02

6 Simon and Joshua's heights are in the ratio 8 : 9. Joshua is 1.71 m tall. How tall is Simon?

Exercise
2.02

7 The lengths of the sides in a right-angled triangle are in the ratio 3 : 4 : 5. The longest side is 30 cm.

   a   Find the lengths of the other two sides.

   b   What is the perimeter of the triangle?

Exercise
2.03

8 Divide each quantity into the given ratio.

   a   20 into 7 : 3   **b**   60 into 3 : 2

   c   120 into 4 : 1   **d**   75 into 2 : 1

   e   50 into 2 : 3   **f**   84 into 4 : 3

Exercise
2.03

9 a Derek needs 24 mL of this shade of red to paint some waratahs in a bush scene. The colour is a 3 : 1 mix of red and yellow. How much of each colour will he need?

   b   This shade of red will show the parts of the waratahs in direct sunlight. It consists of red, yellow and white in the ratio 4 : 1 : 1. Derek requires 12 mL of this colour. How much red, white and yellow paint should he mix together?

Exercise
2.03

10 Asher is making a bracelet using pearls, seed beads and crystals in the ratio 4 : 2 : 3. She uses 54 pieces to make the bracelet. How many of each type will she require?

11 This map of Nambucca Heads has a scale of 1 : 40 000.

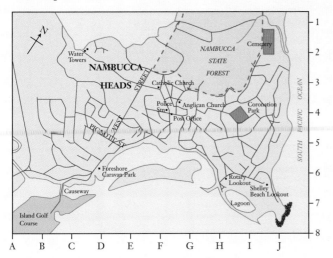

a Find in metres the distance between:

   i the Water Towers (C2) and the Catholic Church (F3)

   ii the Anglican Church (G4) and Coronation Park (I4)

   iii Rotary Lookout (H6) and Shelley Beach Lookout (J6)

   iv the Post Office (F4) and the Foreshore Caravan Park (D6).

b Find the length of:

   i West Street (E4)

   ii Piggott Street (D5).

c How long is the lagoon (I7)?

d What are the dimensions of the cemetery (J2)?

e How long is the causeway leading to the Island Golf Course (B7)?

f To train for a fun run, Tegan decides to run 8 km 3 times a week. What distance will this be on the map? Outline a possible course for her training run, starting and finishing at the Foreshore Caravan Park (D6).

3.

GRAPHING LINES

Chapter problem

Paul owns a factory that makes sails. It costs $9000 per month to cover rent, electricity and wages. Each sail costs $250 to make and sells for $850.

a How many sails does Paul need to make and sell per month to break even?

b What profit or loss will he make in a month if he makes and sells 20 sails?

Alamy Stock Photo/Neil McAllister

CHAPTER OUTLINE

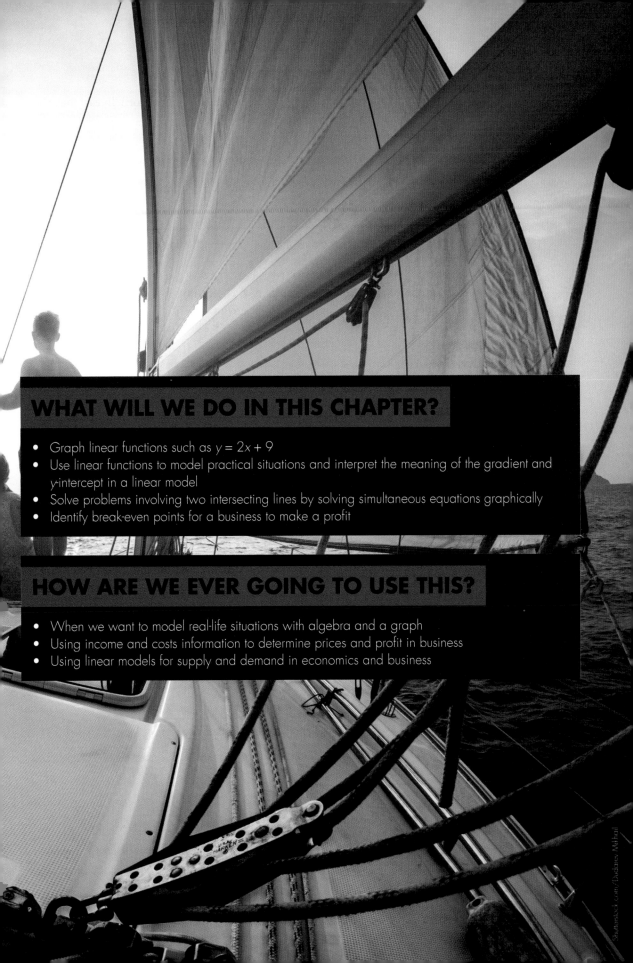

WHAT WILL WE DO IN THIS CHAPTER?

- Graph linear functions such as $y = 2x + 9$
- Use linear functions to model practical situations and interpret the meaning of the gradient and y-intercept in a linear model
- Solve problems involving two intersecting lines by solving simultaneous equations graphically
- Identify break-even points for a business to make a profit

HOW ARE WE EVER GOING TO USE THIS?

- When we want to model real-life situations with algebra and a graph
- Using income and costs information to determine prices and profit in business
- Using linear models for supply and demand in economics and business

3.01 Graphing linear functions

We learned about **linear functions** such as $y = 2x + 9$ in Year 11. The word 'linear' means 'of a line'. The graph of a linear function is a straight line.

> The equation of a line has the form $y = mx + c$, where m is the **gradient** of the line and c is the **y-intercept**.
>
> The **gradient** measures the steepness of a line, and is the number of units the line goes up (or down) for every one unit it goes across to the right.
>
> The **y-intercept** or **vertical intercept** is the value where the line crosses the y-axis.

For example, the graph of $y = 2x + 9$ has a gradient of 2, meaning it goes up two units for every one unit it goes across to the right, and a y-intercept of 9.

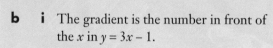

a Draw the graph of the linear function $y = 3x - 1$.

b What is the:

   i gradient?

   ii y-intercept?

Solution

a Complete a table of values for the rule.

x	–2	–1	0	1	2	3
y	–7	–4	–1	2	5	8

Draw a set of axes and plot the points. Rule a straight line through the points, place arrows at each end and label the line with its equation.

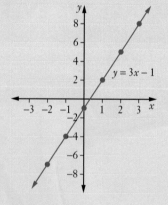

$y = 3x - 1$

b **i** The gradient is the number in front of the x in $y = 3x - 1$.   Gradient is 3

   ii The y-intercept is the constant.   y-intercept is –1

EXAMPLE 2

a What is the gradient of this line?

b What is the value of the *y*-intercept?

c What is the equation of the line?

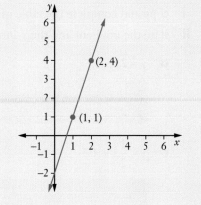

Equation of a line

Gradient and y-intercept of a line

Graphing y = mx + c

Solution

a Draw a right-angled triangle off the 2 points using the line as the hypotenuse.

The line goes up 3 units for every 1 unit across to the right.

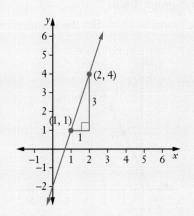

$$m = \frac{3}{1}$$
$$= 3$$

The gradient is 3.

b The *y*-intercept is the value where the line crosses the *y*-axis.

The *y*-intercept is −2.

c The equation of a straight line is $y = mx + c$. From above, $m = 3$ and $c = -2$.

The equation of the line is $y = 3x - 2$.

Exercise 3.01 Graphing linear functions

1 For each linear equation:

   i copy and complete the table of values, then graph the line

  ii state the gradient and the y-intercept of the line.

a $y = 2x - 1$

x	-2	-1	0	1	2
y					

b $y = -2x + 4$

x	-2	-1	0	1	2
y					

c $y = \dfrac{1}{2}x + 1.5$

x	-2	-1	0	1	2
y					

d $y = -2x + 6$

x	-2	-1	0	1	2
y					

2 For each graph, find:

  i the gradient   **ii** the y-intercept   **iii** the equation of the line.

a

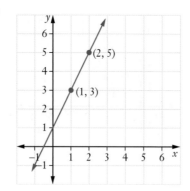

b

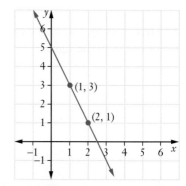

c

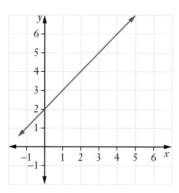

Remember: if the line goes down from left to right, then the gradient is **negative**.

3 For each equation, state:

   i   the gradient   **ii**   the y-intercept.

 Example 1

    a $y = 3x - 5$ **b** $y = -x + 6$

    c $y = 2x + 11$ **d** $y = \dfrac{x}{4} - 5$

Hint: $\dfrac{x}{4}$ is another way of writing $\dfrac{1}{4}x$.

4 Match each linear function to its correct graph. Use technology to check your answers.

    a $y = \dfrac{1}{2}x + 6$ **b** $y = 3x + 6$ **c** $y = \dfrac{1}{2}x + 12$ **d** $y = 2x + 12$

A
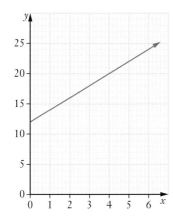

B
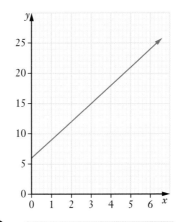

C
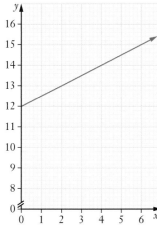

D
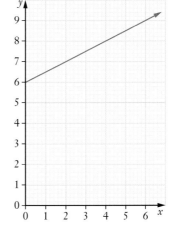

3.02 Linear modelling

In Year 11, we learned that linear functions can be used to model many real-world situations. This is called **linear modelling**, which we will now review.

EXAMPLE 3

Carla has a window-cleaning business. Each day, it costs her an average of $75 for fuel and vehicle maintenance and $8 per job.

Let C = Carla's daily costs in dollars and n = the number of jobs she does per day.

a Write a linear function for Carla's daily costs.

b Construct a graph to illustrate Carla's daily costs.

c What is the gradient and vertical intercept of the graph and what do these values represent?

Solution

a Carla's daily costs
= $75 car costs + $8 per job

$C = 75 + 8 \times n$
$C = 8n + 75$

b Complete a table of values for $C = 8n + 75$.

n	0	1	2	3	4
C	75	83	91	99	107

Use the table of values to graph a line on a number plane.

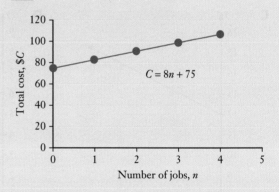

c For $C = 8n + 75$, the gradient is 8 and the vertical intercept is 75.

The gradient represents Carla's costs per job ($8). The vertical intercept represents Carla's fixed daily costs ($75), the costs without doing any jobs.

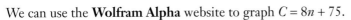

TECHNOLOGY

GRAPHING CALCULATORS AND WEBSITES

We can use technology such as graphics calculators, spreadsheets, graphing software and websites to graph functions.

Wolfram Alpha

We can use the **Wolfram Alpha** website to graph $C = 8n + 75$.

Enter '8n + 75' in the box and press the Enter key (or click '=').

This will output two graphs of $C = 8n + 75$ (for different values of n) and a table of values.

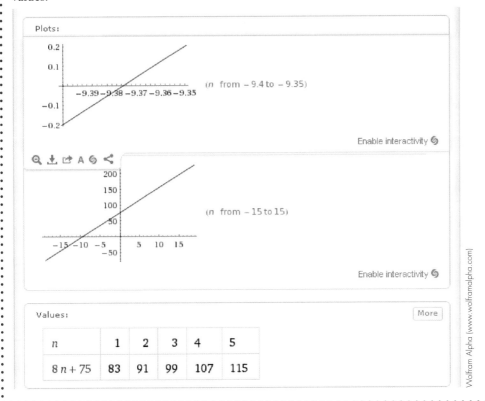

Exercise 3.02 Linear modelling

1 Hawkes Landscaping sells garden soil. The charge is $60 plus $28 per tonne to deliver up to 25 tonnes.

Example
3

a Copy and complete this table of values.

Number of tonnes, n	0	1	2	3	4	5	10	15	20	25
Cost of delivered soil, C										

b Write a linear function for the cost C of n tonnes of soil.

Wolfram Alpha

c Graph this function.

d Use the Wolfram Alpha website or other technology to check your graph.

e What is the vertical intercept of the line?

f What physical quantity does the vertical intercept represent?

g What is the gradient of the line and what does it represent?

h Why do you think the company limits this pricing system to deliveries of up to 25 tonnes?

i If you extrapolated and substituted 40 tonnes of soil into the formula, what value of C do you get?

> When you make a prediction that is beyond or outside the information you have, it's called **extrapolating**, pronounced 'ex-strap-o-late-ing'.

j If you buy 40 tonnes of soil, you will need two deliveries. Calculate the real cost to buy 40 tonnes of soil.

2 Marko sells hot chips at his fast food store. Each day, the cooking oil costs $42 and each cup of chips costs him 75c to make.

a Copy and complete this table.

Number of cups of chips, n	0	10	20	30	40	50	60	100	200
Cost of making chips, $C	42	49.5							

b Write a linear function to calculate the cost of making n cups of chips in one day.

c Use the function to calculate the cost of making 170 cups of chips.

d Yesterday, the cost of making chips was $149.25. How many cups of chips did Marko make?

e Construct a graph showing Marko's daily cost for producing n cups of chips.

f Find the vertical intercept and explain what this value represents.

g Find the gradient and explain what this value represents.

h Dakota's fish-and-chip shop across the road makes chips at only 50c per cup but the daily cost of cooking oil is $48. Describe how the graph of Dakota's costs will be similar but different to Marko's graph. Use technology to check your answer.

3 Belinda's hobby is jumping out of planes! This table of values shows her height above the ground t seconds after she opened her parachute.

Time, t seconds	0	60	90	120
Height, h metres	600	300	150	0

a How high above the ground was Belinda when she opened her parachute?

b How many metres is Belinda falling every second?

c Explain why the linear function $h = 600 - 5t$ represents this situation.

d Graph this function.

4 In Year 11, we learned that new items depreciate over time, and we used the straight-line depreciation method $S = V_0 - Dn$. Eddie bought a new car with a value of $39 000. Suppose it depreciates at $3000 or $4000 per year.

a Copy and complete the table below for two different depreciation amounts per year.

Yearly depreciation	Salvage value, S, after n years				
	1 year	2 years	3 years	4 years	5 years
$3000					
$4000					

b Graph this information on the same axes as 2 straight-line graphs.

c For the depreciation rate of $3000 per year, find:

   i the gradient of the line and explain what this value represents.

   ii the vertical intercept and explain what this value represents.

   iii the equation of the line in the form $S = mn + c$.

d For the depreciation rate of $4000 per year, find:

   i the gradient of the line and explain what this value represents.

   ii the vertical intercept and explain what this value represents.

   iii the equation of the line in the form $S = mn + c$.

e Using the depreciation rate of $3000 per year, how many years will it take for the salvage value to be zero?

f Using the depreciation rate of $4000 per year, after how many years will the car be valued at $11 000?

5 Gabrielle is investing $2000 at a simple interest rate of 3.5% p.a.

   a Copy and complete this table for the interest earned using the formula $I = Prn$.

Number of years, n	0	1	2	3	4	5
Interest earned, $I	0	70				

   b Graph this function.

   c Find the gradient of the line and explain what this value represents.

   d Find the vertical intercept and explain what this value represents.

   e What is the equation of this linear function?

   f Use the equation to find the interest earned if $2000 is invested at 3.5% for 7 years.

   g Gabrielle withdraws her money after 2 years. How much money in total does she receive?

6 The linear function $n = 2.58 + 0.032M$ models the relationship between n, the number of litres of blood in a man 175 cm tall and M, the man's mass in kg. However, this model is valid only for men 175 cm tall with a mass of between 50 and 200 kg.

   a Graph the linear function $n = 2.58 + 0.032M$.

   b Use the function to calculate the number of litres of blood in the body of a 10 kg boy.

   c Give 2 reasons why your answer to part **b** is invalid.

   d Derek is 175 cm tall and lost 80 kg on the TV show *The Biggest Loser*. His initial body mass was 175 kg. By how many litres did his blood volume reduce?

3.03 Intersecting lines

Simultaneous equations

Intersection of
lines

A pair of equations, such as $y = x + 2$ and $y = 2x - 3$, can be solved together to find the values of x and y that satisfy *both* equations. Because they are solved at the same time, they are called **simultaneous equations**.

When you graph simultaneous equations on a number plane, their solution are the coordinates of the point where their graphs intersect.

ISBN 9780170413534

a Graph the equations $y = x + 2$ and $y = 2x - 3$ on the same axes.

b Find the coordinates of the point of intersection of the 2 lines.

c Write the solution of the simultaneous equations $y = x + 2$ and $y = 2x - 3$.

Solution

a Graph the 2 lines by completing a table of values for each equation and plotting the points on the number plane.

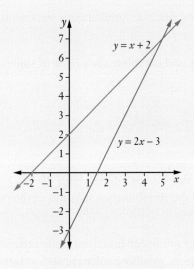

b Read the point of intersection from the graph.

The point of intersection is $(5, 7)$.

c Use the coordinates of the point of intersection to write the solution.

The solution to the simultaneous equations is: $x = 5, y = 7$

Note: We can check this solution by calculating y when $x = 5$.

For $y = x + 2$:   For $y = 2x - 3$:
  $y = 5 + 2$ $y = 2(5) - 3$
    $= 7$ ✓ $= 7$ ✓

Exercise 3.03 Intersecting lines

1 a Copy and complete each table of values.

$y = 2x - 1$

x	0	1	3
y			

$y = 5 - x$

x	0	3	5
y			

b Graph both lines on the same axes.

c Hence, find the simultaneous solution to the pair of equations $y = 2x - 1$ and $y = 5 - x$.

2 a Copy and complete each table of values.

$x + y = 4$

x	0	2	4
y			

$2x - y = 5$

x	0	2	4
y			

b Graph both lines on the same axes.

c Find the simultaneous solution of the pair of equations $x + y = 4$ and $2x - y = 5$.

3 We can graph linear functions using technology such as graphics calculators, spreadsheets, graphing software and websites. Use the technology of your choice to graph and solve each pair of simultaneous equations.

a $y = 10 - 2x$ and $y = 3x$

b $y = 2x + 7$ and $y = 1 - x$

c $p = 2q + 11$ and $p = 2 - q$

d $y = 2 - x$ and $y = 3x - 10$

e $y = x - 4$ and $y = 3x + 2$

f $y = 4x$ and $y = -2x + 6$

ISBN 9780170413534

3.04 Intersecting lines in modelling

Simultaneous equations can be used to solve problems involving linear modelling.

EXAMPLE 5

Cherie sells scones for $3 each. It costs her $240 for the necessary equipment and $1 to make each scone.

a Find a linear function for C, the cost in dollars for making n scones.

b Find a linear function for I, the income in dollars from selling n scones.

c Graph both functions on the same axes for values of n from 0 to 200.

d How many scones does Cherie need to sell to break even?

e How much profit will Cherie make if she sells 390 scones?

Solution

a The cost is $240 plus $1 for each scone Cherie makes.

$$\text{Cost} = \$240 + n \times \$1$$
$$C = 240 + n$$
$$C = n + 240$$

b Income is $3 for each scone.

$$\text{Income} = n \times \$3$$
$$I = 3n$$

c Graph both lines together, either by hand or using technology.

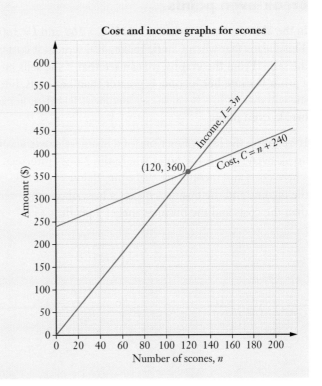

Cost and income graphs for scones

d Break even means that costs and income are the same. This is the point where the lines intersect. The lines intersect at the point (120, 360), where 120 represents the number of scones.

Cherie needs to sell 120 scones to break even.

e The profit is how much the income is more than the cost: substitute $n = 390$ into both the Income and Cost functions.

Income: $I = 3n$
$= 3 \times 390$
$= \$1170.$
For 390 scones, the income will be $1170.
Cost: $C = n + 240$
$= 390 + 240$
$= 630$
For making 390 scones, the cost will be $630.

Profit = income − cost

Profit = $1170 − $630
$= \$540$
Cherie will make a $540 profit when she makes and sells 390 scones.

Break-even points

In the above example, when $n = 120$, $C = 360$ and $I = 360$. The cost and income are the same. This means that when Cherie makes 120 scones, it costs her $360 and she receives $360 from the sales. From the graphs, we can see that if she sells more than 120 scones, she will make a profit because her income is greater than her costs, but if she sells fewer than 120 scones, she will make a loss because her income is less than her costs. The value of 120 is called the **break-even point**.

If we graph the linear functions of a business' **costs** and **income** (or **revenue**), then the break-even point is the point where the 2 lines intersect. The break-even point is the position where costs and income are the same. Below the break-even point, the costs are greater than the income and the business makes a loss. Above the break-even point, the income is greater than the costs and the business makes a profit.

ISBN 9780170413534

Exercise 3.04 Intersecting lines in modelling

1 Khalid bakes and sells muffins. The equipment used to bake the muffins costs $200 and each muffin he makes costs $1. Khalid sells the muffins for $2 each.

Example
5

Let n = the number of muffins, C = the total cost to make n muffins and I = the income from selling n muffins.

a Explain why $C = n + 200$ and $I = 2n$.

b Graph $C = n + 200$ and $I = 2n$ on the same axes.

c What are the coordinates of the point where the two lines intersect?

d What does this point represent?

e Will Khalid make a profit if he sells 120 muffins? Explain your answer.

f How much profit will Khalid make if he sells 230 muffins?

2 On the Easter weekend, Dane and Mitchell headed north in different cars to go camping. Dane travelled at 60 km/h. Mitchell left an hour later and took a different route, travelling at 90 km/h. The graph shows the distances they have each travelled t hours after Dane left town.

a D is the distance travelled from town during t hours. Explain why the formula $D = 60t$ represents the distance Dane is from town after t hours.

b The distance Mitchell travelled out of town is represented by $D = 90(t - 1)$. What is the solution to the simultaneous equations $D = 60t$ and $D = 90(t - 1)$?

c What does this solution represent?

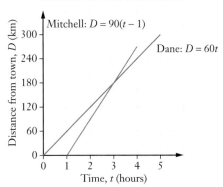

Distance of Dane and Mitchell

3 Chloe is deciding between two mobile phone plans. Plan A offers her no connection fee and charges of 40c per minute for each call. Calls under plan B cost 20c per minute with a 60c connection fee. The graphs show the costs of phone calls under each plan for calls of up to 6 minutes.

a Which plan (A or B) has the equation $C = 0.6 + 0.2m$? Explain the reason for your choice.

b For what length of phone call is the cost the same under both plans?

c Which plan is the cheaper for longer calls?

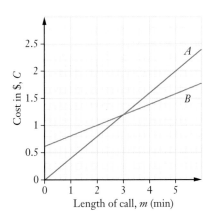

4 A truck and a car were travelling on the freeway. When an accident occurred ahead of them, both vehicles had to stop quickly. The graphs show the speed each vehicle was travelling at t seconds after they each applied their brakes.

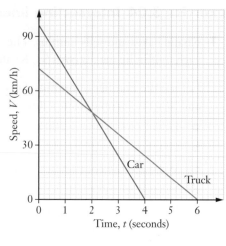

a How fast were the car and the truck travelling in km/h before the accident?

b How many seconds after applying their brakes were the vehicles travelling at the same speed?

c How much longer than the car did it take for the truck to stop?

d Write a linear equation for V, the speed in km/h, for the speed of each vehicle t seconds after applying the brakes.

e Explain why the gradients of both lines are negative.

f Suggest a reason why the truck took longer to stop than the car, even though the car was travelling faster initially.

5 Allana sells flowers at the market for $10 per bunch.

a Copy and complete this table.

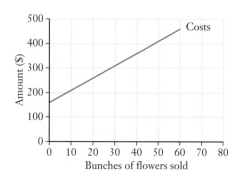

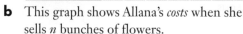

Bunches of flowers sold	0	10	20	30	40	50	60
Revenue received, $							

b This graph shows Allana's *costs* when she sells n bunches of flowers.

Copy the graph and use your answers to part **a** to graph Allana's revenue on the same axes.

c Explain how you know that Allana will break even when she sells 32 bunches of flowers.

d How many bunches of flowers will Allana need to sell to make a profit?

ISBN 9780170413534

6 Each week, a toy factory's fixed expenses (costs) total $1500. Its expenses increase by $15 for every toy it produces. Each item produced sells for $35.

Let n = the number of items the factory produces each week, C = the total costs to produce n items, and I = the income the factory receives from selling n items.

a Does the expression $35n$ represent the income from selling n items or the cost to produce n items?

b What does the expression $15n + 1500$ represent?

c Graph the linear functions $I = 35n$ and $C = 1500 + 15n$ together.

d How many items does the factory need to produce each week to break even?

e How much profit will the company make if it sells 100 items in a week?

7 Jon lives near the snowfields and earns money taking tourists on dog sled rides in the snow. Jon has displayed his expenses and income on a graph.

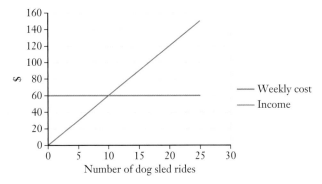

a How much are Jon's weekly costs?

b How many dog sled rides does Jon require each week to break even?

c How much does Jon charge for a dog sled ride?

d Suggest a reason why Jon's cost line is horizontal.

8 To advertise the school musical, the drama class is making and selling promotional T-shirts. The set-up costs to make the T-shirts is $160 and each shirt will cost $10 to make. The class will sell the T-shirts for $20 each.

a Find a linear function for C, the cost in dollars for making n T-shirts.

b Find a linear function for I, the income in dollars from selling n T-shirts.

c Graph both functions on the same axes for values of n from 0 to 20.

d How many T-shirts need to be sold to break even?

e The class estimates it will sell 90 T-shirts. How much profit can they expect to make?

9 Grant is hard of hearing and uses his mobile phone for text messages only. The graph shows the monthly cost, D dollars for n text messages on 2 phone plans. Plan A is a fixed price per month with unlimited messages, whereas Plan B has a monthly charge and a cost per message.

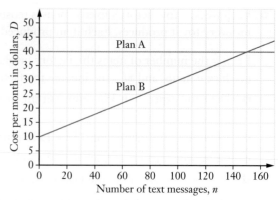

a What is the monthly cost for Plan A?

b What is the vertical intercept of the graph of Plan B?

c What does the vertical intercept in Plan B represent?

d How much does each message cost to send on Plan B?

e What is the equation of the line representing Plan B?

f For how many texts per month is Plan A the same cost as Plan B?

g Grant sends an average of 5 or 6 text messages per day. Which plan would you recommend he chooses? Use calculations to justify your recommendation.

INVESTIGATION

MODELLING THE YEAR 12 FORMAL

In this investigation we will model the costs and income for running the Year 12 formal.

1 You will need to collect the following information for where you live:

Costs	Hire of venue	_____
	Decorations	_____
	Band/DJ	_____
	Other costs	_____
	Total costs	_____

These are the fixed costs for the formal.

The committee decides to charge $15 per person to cover the fixed costs.

2 On the same number plane:

- draw a graph for $I = 15n$ for n from 0 to 200.

- draw a graph for $F = \underline{\hspace{1.5cm}}$ for the fixed costs. This will be a horizontal line.

3 a Where do the two lines intersect? How many people need to attend the formal to cover the fixed costs?

b How many people do you expect to attend the formal? Will you cover the fixed costs if you charge $15 per person?

c Depending on what you have found, adjust the amount you need to charge per person to cover the fixed costs. This could be more or less than $15. Draw the income graph for this new amount.

d How many people now need to attend the formal to cover the fixed costs?

4 Find out from the venue how much they charge per person for catering and calculate what you need to charge for your formal tickets.

KEYWORD ACTIVITY

COMPLETE THE BLANKS

Use the listed words below to copy and complete a summary of this chapter.
Some of the words are used more than once.

break-even	costs	depreciation	gradient
income	interest	intersecting	linear
loss	model	profit	simultaneous
straight	*y*-intercept		

In this chapter, we reviewed what we learned in Year 11 about **1**_____ functions. These functions can be graphed as **2**_____ lines. The **3**_____ of the line tells us how steep the line is. The **4**_____ tells us where the straight line crosses the *y*-axis.

We used straight-line graphs to **5**_____ real-life contexts such as business **6**_____, simple **7**_____ and **8**_____.

We learned that **9**_____ lines can be used to solve **10**_____ equations. We applied this idea to practical problems such as business, travelling and comparing mobile phone plans.

In business, we draw a graph for the money coming in, called **11**_____, and the money being spent, called **12**_____. The point of intersection of these two lines is called the **13**_____ point. At this point, the business owner does not make a **14**_____ or a **15**_____.

SOLUTION TO THE CHAPTER PROBLEM

Problem

Paul owns a factory that makes sails. It costs $9000 per month to cover rent, electricity and wages. Each sail costs $250 to make and sells for $850.

a How many sails does Paul need to make and sell per month to break even?

b What profit or loss will he make in a month if he makes and sells 20 sails?

Solution

Let n = the number of sails, C = the cost to produce n sails per month, and I = the income from selling n sails.

a The linear functions are:

$C = 250n + 9000$

$I = 850n$

The lines intersect at (15, 12 750), which means the break-even point is $n = 15$.

Paul needs to sell 15 sails per month to break even.

b When $n = 20$:

$\text{Cost} = \$250 \times 20 + \$9000$

$= \$14\,000$

$\text{Income} = \$850 \times 20$

$= \$17\,000$

Income is greater than cost, so a profit will be made.

$\text{Profit} = \$17\,000 - \$14\,000$

$= \$3000$

The factory will make $3000 profit when it makes and sells 20 sails in a month.

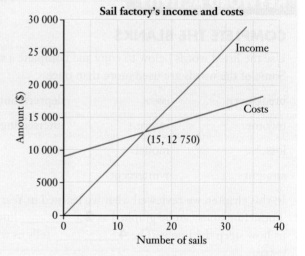

Sail factory's income and costs

3. CHAPTER SUMMARY

- Give an example of a context that could be modelled using 2 linear functions.

- Explain what the point of intersection would mean for your example.

- Is there any part of the topic you didn't understand? If so, ask your teacher for help.

Copy and complete this mind map of the topic, adding detail to its branches and using pictures, symbols and colour where needed. Ask your teacher to check your work.

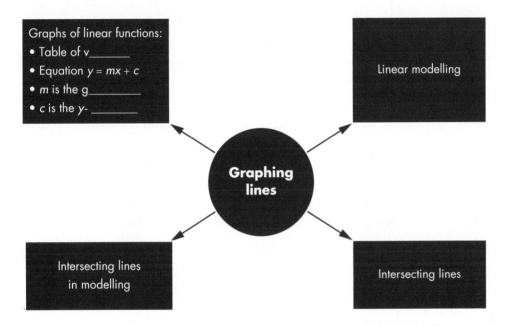

Graphs of linear functions:
- Table of v_____
- Equation $y = mx + c$
- m is the g_____
- c is the y-_____

Linear modelling

Graphing lines

Intersecting lines
in modelling

Intersecting lines

Exercise
3.01

1 Graph each linear function and state its gradient and y-intercept.

    a $y = x + 3$ **b** $y = -2x - 1$

Exercise
3.01

2 For each graph find:

    i the gradient   **ii** the y-intercept   **iii** the equation of the line.

a

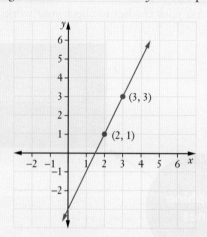

(3, 3)

(2, 1)

b

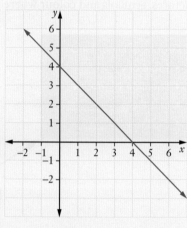

Exercise
3.02

3 Adrian's Awesome Appetisers charges $10 per person plus an $80 fixed charge.

    a Copy and complete this table.

Number of people, n	0	50	100	150	200	250
Charge, C	80					

    b Construct a graph showing Adrian's charges for n people.

    c Write a linear function to calculate the charge for n people.

    d Use the function to calculate the cost for 170 people.

    e Adrian charged NCM Properties $830 for catering a party.
       How many people were catered for?

    f Find the vertical intercept of the line and explain what this value represents.

    g Find the gradient and explain what this value represents.

Exercise
3.03

4 Use the technology of your choice to graph and solve each pair of simultaneous equations.

    a $y = x + 4$ and $y = 2x - 3$

    b $y = -x + 2$ and $y = 3x + 6$

5 The Year 12 fundraising committee plans to sell school souvenir USB drives. The manufacturer quoted an 'initial charge of $126 plus $5 per unit' and the committee thinks it will sell the units for $12 each.

Exercise
3.04

   a Find a linear function for $C, the cost of making n USB drives.

   b Find a linear function for $I, the income from selling n USB drives.

   c Graph both functions on the same axes for values of n from 0 to 20.

   d How many USB drives does the committee need to sell to break even?

   e There are 140 students in Year 12. If half of them buy a USB drive each, how much profit can the committee expect to make?

FINANCIAL MATHEMATICS

4.

DEPRECIATION
AND LOANS

Chapter problem

Leslie borrowed $240 000 to buy an apartment in which to live. She is going to repay the loan plus interest in monthly instalments at 7.8% p.a. monthly reducible finance.

If Leslie borrows the money over 15 years, the monthly repayments will be $2265.95 and the repayments for the loan over 30 years will be $1727.70.

a Calculate the total amount Leslie will repay if she takes the loan for 15 years or for 30 years.
b Suggest a reason why Leslie decided to take the loan for 30 years.

CHAPTER OUTLINE

Note: Declining-balance depreciation and reducing balance loan graphs are covered in chapter 6.

WHAT WILL WE DO IN THIS CHAPTER?

- Calculate salvage value and depreciation using the straight-line and declining-balance methods of depreciation
- Use technology to examine the progress of a reducing balance loan
- Investigate the effect of the interest rate and repayment amount on the time taken to repay a loan
- Investigate situations and solve problems involving credit cards

HOW ARE WE EVER GOING TO US THIS?

- When calculating depreciation of an asset as a tax deduction for income tax or company tax
- When calculating the total amount we will repay for a loan and the included interest charges
- When borrowing money to buy a car or home
- When spending money at a store or online using a credit card

4.01 Straight-line depreciation

As soon as we purchase new assets such as cars, work equipment and furniture, their value decreases. For new cars, the drop in value in the first few years can be very substantial, although it is less severe in later years. In Year 11, we learned that the decrease in an item's value is called **depreciation**. The value of the item at any time is called its **salvage value**.

If you own a car, computer, office equipment or tools and you use them in your work, you can claim the depreciation of these items as an **allowable tax deduction**.

In **straight-line depreciation**, we decrease the value of the asset by the same amount each year. If we graph the value of the item as it gets older, the graph is a straight line.

The straight-line depreciation formula

$$S = V_0 - Dn$$

where S is the salvage (current) value of the asset

V_0 is the initial value

D is the amount of depreciation per period

n is the number of periods.

EXAMPLE 1

Jamie bought office furniture for $17 000. It is depreciating at a rate of 10% p.a. of its original value each year.

a What was its salvage value after 4 years?

b What was the total depreciation over the 4 years?

Solution

a In the formula $S = V_0 - Dn$,
$V_0 = \$17\ 000$, $D = 10\% \times \$17\ 000$,
$n = 4$.

Annual depreciation, $D = 10\% \times \$17\ 000$
$= \$1700$

$S = V_0 - Dn$
$= \$17\ 000 - \$1700 \times 4$
$= \$10\ 200$

Write the answer.

The salvage value after 4 years is $10 200.

b Total depreciation is annual
depreciation $\times 4$

Total depreciation $= \$1700 \times 4$
$= \$6800$

OR original value – salvage value

Total depreciation $= \$17\ 000 - \$10\ 200$
$= \$6800$

EXAMPLE 2

Tania bought a used car for $22 500. It is expected to depreciate by $2400 each year.

a By how much will the car depreciate over the next 3 years?

b What will be its salvage value after 3 years?

Solution

a Total depreciation is annual depreciation × 3.

Depreciation = $2400 × 3

$= 7200

b Subtract the depreciation from the original value.

Salvage value = original value − depreciation

$= $22 500 − 7200

$= $15 300$

Exercise 4.01 Straight-line depreciation

In this exercise, round all answers to the nearest cent where appropriate.

1 Lena has bought a new SUV for $55 000. She has been told that she can expect it to depreciate 10% of the original price each year. Calculate:

a the car's salvage value after 4 years

b the amount it has depreciated in that time.

2 Melissa bought a new home gym for $2400. The gym is expected to depreciate by $350 each year.

a By how much will the gym depreciate in 5 years?

b Calculate the gym's salvage value at the end of 5 years.

3 Yusef believes that his new car, which he purchased for $42 000, will depreciate by $3500 per year. Calculate the car's salvage value after 5 years.

4 Liam's second-hand Mini Coupe cost him $32 000. It is expected to depreciate at a constant rate of 12% of the original price each year for the next 4 years.

a How much will the car depreciate each year?

b How much will Liam's car depreciate in 4 years?

c Calculate the car's salvage value after 4 years.

5 Kristy bought a new lounge suite for $5200. She expects the lounge suite to lose 25% of its original price in the first year and 10% of the original price each year after that.

 a How much value will the lounge suite lose in the first year?

 b How much value will the lounge suite lose in the second year?

 c Calculate the salvage value of Kristy's lounge suite after 4 years.

6 Samantha bought a new computer system for her business for $6420. Each year, she is going to claim a $1605 tax deduction for depreciation.

 a Calculate the salvage value of the computer system when it is 1 year old.

 b What percentage rate of depreciation is Samantha using?

 c When will the salvage value of the computer system be zero?

7 Tony bought some new tools to use in his work. During the first year, the tools depreciated by $275 and their salvage value was $840. How much were Tony's tools worth when they were new?

8 Nicky has a dog wash van that she uses in her mobile dog-washing business. She paid $25 000 for the van and she is claiming $5000 depreciation annually on her tax.

 a Calculate the salvage value of the van when it is 3 years old.

 b When will the salvage value of the van be zero?

ISBN 9780170413534

4.02 Declining-balance depreciation

With **straight-line depreciation**, the value of the item decreases by the same dollar amount each year.

With **declining-balance depreciation**, the value of the item decreases by a *percentage* of the current value of the item, so it is not constant.

Declining-balance depreciation: spreadsheet

Declining-balance depreciation formula practice

The declining-balance depreciation formula

$$S = V_0 (1 - r)^n$$

where S is the salvage (current) value of the asset

V_0 is the initial value

r is the depreciation rate per time period, expressed as a decimal

n is the number of periods.

The declining-balance depreciation formula is similar to the compound interest formula $FV = PV(1 + r)^n$, except that the '+' sign is replaced by a '−' sign to indicate something decreasing by a percentage rather than increasing.

EXAMPLE 3

A car purchased for $32 000 depreciates at 8% per year.

a   What is its salvage value after 6 years?

b   What was the depreciation during that time?

Declining-balance depreciation

Solution

a $S = V_0 (1 - r)^n$

where $V_0 = \$32\ 000, r = 0.08, n = 6$

$S = \$32\ 000 \times (1 - 0.08)^6$
$= \$32\ 000 \times (0.92)^6$
$= \$19\ 403.36004$
$\approx \$19\ 403.36$

b   Depreciation
= original value − salvage value

Depreciation
$= \$32\ 000 - \$19\ 403.36$
$= \$12\ 596.64$

Write the answer.

During the 6 years, the car's value decreased by $12 596.64.

EXAMPLE 4

Rhonda is a graphic designer who works from home. Her new computer cost $5400 and she uses a declining-balance rate of depreciation of 22% to claim as a tax deduction on the computer.

Age in years	Salvage value
0	$5400
1	$4212
2	$3285
3	$2563

Use the table to calculate the tax deduction Rhonda can claim when her computer is:

a 1 year old

b 2 years old.

Solution

a Tax deduction after 1 year   Depreciation, 1st year = $5400 – $4212
     = the depreciation in the first year.   = $1188

b Tax deduction after 2 years   Depreciation, 2nd year = $4212 – $3285
     = the depreciation in the second year.   = $927

Exercise 4.02 Declining-balance depreciation

1 Use the formula $S = V_0 (1 - r)^n$ to calculate the value of S when $n = 8$, $r = 7\%$ and $V_0 = \$2000$. Answer correct to the nearest dollar.

2 A new office printer cost $625 and it is depreciating at 20% per year.

  a Calculate the printer's salvage value when it is 3 years old.

  b How much will the printer depreciate during its first 3 years?

3 Jakub's Peugeot cost him $22 000. It will depreciate at 12% each year.

  a Calculate its value after 7 years, correct to the nearest dollar.

  b By how much will it depreciate over the 7 years?

4 This table shows the salvage value of an industrial carpet-cleaning machine using declining-balance depreciation.

Example
4

Age in years	Salvage value to the nearest dollar
0	$18 000
1	$15 300
2	$13 005
3	$11 054

a How much was the machine worth when it was new?

b By how much did the machine depreciate in the first year?

c How much did the machine depreciate during the 3rd year?

d What is the percentage decrease (rate of depreciation) of the machine each year?

5 Elena can claim a tax deduction for the depreciation of items in her investment property. The new carpet cost $3200 and Elena is depreciating it at 10% p.a.

a Use the formula to calculate the carpet's salvage value at the end of the 3rd and 4th years, correct to the nearest dollar.

b Hence, calculate the value of the carpet's depreciation during the 4th year.

c Calculate the carpet's salvage value when it is 5 years old.

d Hence how much can Elena claim as a tax deduction for depreciation of her carpet during the 5th year?

6 The declining-balance depreciation on a photocopier is 25% p.a. After 5 years, the salvage value of the photocopier was $925.

a Explain why the solution to the equation $925 = V_0 \times (1 - 0.25)^5$ gives the photocopier's new value.

b Calculate the photocopier's new value, correct to the nearest $10.

7 Thieves stole some of Robert's photographic equipment. Robert paid $15 000 for the equipment when it was new and when it was 2 years old it had a salvage value of $8438.

a Show that the solution to the equation $8438 = 15\,000 \times (1 - r)^2$ represents the annual rate of declining-balance depreciation.

b Calculate the rate of declining-balance depreciation as a percentage, correct to the nearest percentage.

8 This table shows the salvage value of a $6000 machine using declining-balance and straight-line depreciation.

Time in years	Salvage value Straight-line depreciation	Salvage value Declining-balance depreciation
0	$6000	$6000
1	$4800	$3600
2	$3600	$2160
3	$2400	$1296
4	$1200	$778
5	0	$467

a By how much is the machine depreciating each year with straight-line depreciation?

b Complete the table of values to show the declining-balance depreciation each year.

Year	1st	2nd	3rd	4th	5th
Depreciation					

c Which method gives the greater depreciation for the first 2 years?

d Which method gives the greater total depreciation for 5 years?

9 Ask your teacher to download the Declining-balance depreciation spreadsheet from NelsonNet. What you have to do:

Declining-balance depreciation

- Systematically change the size of the rate of depreciation. Try increasing values such as 2%, 5%, 10%, 20%, 30% and observe how the graphs change.

- Describe the effect on the graphs of increasing the annual rate of declining-balance depreciation.

4.03 Reducing balance loans

Reducing balance loan grids

When we borrow money from a financial institution such as a bank, credit union or finance company, we are charged interest. The interest rate depends on the type of financial institution and the reason we are borrowing the money. Wise consumers shop around for the best value before they commit themselves.

There are two types of interest on loans. With **simple interest**, also known as **flat-rate interest**, we pay the same amount of interest throughout the loan, regardless of the amount we owe.

With **reducible interest**, we pay interest only on the amount of money still owing. As the amount we owe decreases, the interest decreases.

This table shows the progress of a **reducing balance loan**. The principal is $15 000, borrowed at 9% p.a. monthly reducible interest, with monthly repayments of $800. All amounts are rounded to the nearest cent.

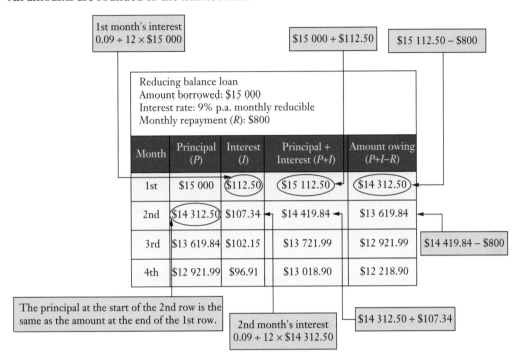

1st month's interest
$0.09 \div 12 \times \$15\,000$

$\$15\,000 + \$112.50$

$\$15\,112.50 - \$800$

Reducing balance loan
Amount borrowed: $15 000
Interest rate: 9% p.a. monthly reducible
Monthly repayment (R): $800

Month	Principal (P)	Interest (I)	Principal + Interest (P+I)	Amount owing (P+I−R)
1st	$15 000	$112.50	$15 112.50	$14 312.50
2nd	$14 312.50	$107.34	$14 419.84	$13 619.84
3rd	$13 619.84	$102.15	$13 721.99	$12 921.99
4th	$12 921.99	$96.91	$13 018.90	$12 218.90

$\$14\,419.84 - \$800$

The principal at the start of the 2nd row is the same as the amount at the end of the 1st row.

2nd month's interest
$0.09 \div 12 \times \$14\,312.50$

$\$14\,312.50 + \$107.34$

4. Depreciation and loans

EXAMPLE 5

When Ethan borrowed $10 000 from the bank to buy a small car, the bank charged
9% p.a. reducible interest and his monthly repayments were $240. Find the values of **a** to
f in the table, then find how much Ethan will owe after he has made 3 repayments and the
total interest he will pay in the first 3 months. Round all values to the nearest cent.

Ethan's reducing balance car loan				
Amount borrowed: $10 000				
Interest rate: 9% p.a. monthly reducible				
Monthly repayments: $240				
Month	Principal (P)	Interest (I)	Principal + interest ($P + I$)	Amount owing ($P + I - R$)
1st	$10 000	$75	$10 075	$9835
2nd	a	b	c	d
3rd	e	f	g	h

Solution

a The principal for the 2nd month is the same as the amount owing at the end of the 1st month.

$9835

b Calculate 1 month's interest at 9% p.a. on the amount Ethan owes.

$9835 × 0.09 ÷ 12 = $73.76

c $P + I$ is the sum of **a** and **b**.

$9835 + $73.76 = $9908.76

d The amount owing at the end of the 2nd month is **c** minus the monthly repayment of $240.

$9908.76 − $240 = $9668.76

e The principal for the 3rd month is the same as the amount owing at the end of the 2nd month.

$9668.76

f Calculate 1 month's interest on $9668.76.

$9668.76 × 0.09 ÷ 12 = $72.52

g $P + I$ is the sum of **e** and **f**.

$9668.78 + $72.52 = $9741.28

h The amount owing at the end of the 3rd month is **g** minus $240.

$9741.28 − $240 = $9501.28

Ethan's reducing balance car loan

Amount borrowed: $10 000

Interest rate: 9% p.a. monthly reducible

Monthly repayments: $240

Month	Principal (P)	Interest (I)	Principal + interest (P + I)	Amount owing (P + I − R)
1st	$10 000	$75	$10 075	$9835
2nd	$9835	$73.76	$9908.76	$9668.76
3rd	$9668.76	$72.52	$9471.28	$9501.28

The amount Ethan still owes after he has made 3 repayments is the amount owing after the 3rd month.

$9501.28

The interest Ethan has paid is the sum of the values in the interest column.

Interest = $75 + $73.76 + $72.52
= $221.28

Exercise 4.03 Reducing balance loans

1 Chantelle borrowed $7000 from the bank at 9% p.a. monthly reducible interest to buy her first car. Her monthly repayments are $320.

Example 5

  a Show that Chantelle will pay $52.50 interest in the first month.

  b Show that Chantelle will owe $7052.50 at the end of the first month immediately before she makes her first repayment.

  c Explain how Chantelle will owe $6732.50 immediately after she makes her first repayment.

  d Copy and complete the next 2 rows of Chantelle's repayment table.

Chantelle's reducing balance car loan

Amount borrowed: $7000

Interest rate: 9% p.a. monthly reducible

Monthly repayments: $320

Month	Principal (P)	Interest (I)	Principal + interest (P + I)	Amount owing (P + I − R)
1st	$7000	$52.50	$7052.50	$6752.50
2nd	i	ii	iii	iv
3rd	v	vi	vii	viii

  e How much will Chantelle owe immediately after she makes her 3rd repayment?

  f How much interest will Chantelle pay in the first 3 months of the loan?

  g How much less interest will she pay in the 4th month than in the 1st month?

  h Why is the amount of interest getting smaller each month?

2 a Copy and complete the first 4 lines of Bianca's personal loan.

Bianca's reducing balance personal loan				
Amount borrowed: $16 000				
Interest rate: 7.2% p.a. monthly reducible				
Monthly repayments: $400				
Month	**Principal** (P)	**Interest** (I)	**Principal + interest** $(P + I)$	**Amount owing** $(P + I - R)$
1st	$16 000	$96	$16 096	$15 696
2nd				
3rd				
4th				

b How much will Bianca owe immediately after she has made her 4th repayment?

c What percentage of Bianca's 4th repayment was interest? Give your answer correct to one decimal place.

3 Ryan borrowed $18 000 at 8% p.a. reducible interest, calculated every 6 months, to expand his business. He agreed to repay the loan over 3 years with half-yearly repayments of $3434.

a Calculate the amount of interest that Ryan will pay after the first half-year.

b Copy and complete the missing values in Ryan's repayment table.

Ryan's business loan				
Amount borrowed: $18 000				
Interest rate: 8% p.a. reducible				
6-monthly repayments: $3434				
Half-years	**Principal** (P)	**Interest** (I)	**Principal + interest** $(P + I)$	**Amount owing** $(P + I - R)$
1st	$18 000	$720	$18 720	$15 286
2nd	**i**	**ii**	$15 897.44	**iii**
3rd	$12 463.44	**iv**	$12 961.98	$9527.98
4th	$9527.98	$381.12	$9909.10	$6475.10
5th	$6475.10	$259.00	**v**	$3300.10
6th	$3300.10	$132.00	$3432.10	0

c After how long will Ryan have repaid more than half the loan?

ISBN 9780170413534

4 Jackson borrowed $12 800 at 7.56% p.a. monthly reducible interest to buy a boat.

a Copy and complete the first 3 rows of this table.

Jackson's loan				
Amount borrowed: $12 800				
Interest rate: 7.56% p.a. monthly reducible				
Monthly repayments: $900				
Month	Principal (P)	Interest (I)	Principal + interest (P + I)	Amount owing (P + I − R)
1st	$12 800			
2nd				
3rd				

b Calculate the total amount Jackson will repay in the first 3 months.

c How much will Jackson repay off the principal in the first 3 months of the loan?

d How much interest will he pay in the first 3 months?

4.04 Loan spreadsheets

Spreadsheets are a useful tool for completing the financial calculations involved in a reducing balance loan.

Reducing balance loan: spreadsheet

Reducible loans

EXAMPLE 6

To answer this question, ask your teacher to download the Reducible loans spreadsheet from NelsonNet. Jordan is borrowing $20 000 at 6% p.a. reducing interest and his monthly repayments are $810.

a How long will it take Jordan to repay the loan?

b How much interest will Jordan pay?

Solution

In the blue cells, enter 20 000 for the loan amount, 0.06 for the interest rate and 810 for his monthly repayment. Remember NOT to put any spaces, commas or $ signs when you type in the money values.

Reducible loans

Only enter data in cells shaded blue. Enter the loan and repayment amounts without any spaces, commas or $ sign.

	Loan amount	$20,000.00
	Interest rate as a decimal	0.06
	Monthly repayment	$810.00

	Amount owing at the beginning of the month	Interest charge for the month	Amount owing plus interest	Amount owing after the repayment
1st month	$20,000.00	$100.00	$20,100.00	$19,290.00
2nd month	$19,290.00	$96.45	$19,386.45	$18,576.45
3rd month	$18,576.45	$92.88	$18,669.33	$17,859.33
4th month	$17,859.33	$89.30	$17,948.63	$17,138.63
5th month	$17,138.63	$85.69	$17,224.32	$16,414.32
6th month	$16,414.32	$82.07	$16,496.39	$15,686.39
7th month	$15,686.39	$78.43	$15,764.83	$14,954.83
8th month	$14,954.83	$74.77	$15,029.60	$14,219.60
9th month	$14,219.60	$71.10	$14,290.70	$13,480.70
10th month	$13,480.70	$67.40	$13,548.10	$12,738.10
11th month	$12,738.10	$63.69	$12,801.79	$11,991.79
12th month	$11,991.79	$59.96	$12,051.75	$11,241.75
13th month	$11,241.75	$56.21	$11,297.96	$10,487.96
14th month	$10,487.96	$52.44	$10,540.40	$9,730.40
15th month	$9,730.40	$48.65	$9,779.05	$8,969.05
16th month	$8,969.05	$44.85	$9,013.90	$8,203.90
17th month	$8,203.90	$41.02	$8,244.92	$7,434.92
18th month	$7,434.92	$37.17	$7,472.09	$6,662.09
19th month	$6,662.09	$33.31	$6,695.40	$5,885.40
20th month	$5,885.40	$29.43	$5,914.83	$5,104.83
21st month	$5,104.83	$25.52	$5,130.35	$4,320.35
22nd month	$4,320.35	$21.60	$4,341.95	$3,531.95
23rd month	$3,531.95	$17.66	$3,549.61	$2,739.61
24th month	$2,739.61	$13.70	$2,753.31	$1,943.31
25th month	$1,943.31	$9.72	$1,953.03	$1,143.03
26th month	$1,143.03	$5.72	$1,148.74	$338.74
27th month	$338.74	$1.69	$340.44	-$469.56
28th month	-$469.56	-$2.35	-$471.91	-$1,281.91
29th month	-$1,281.91	-$6.41	-$1,288.32	-$2,098.32
30th month	-$2,098.32	-$10.49	-$2,108.81	-$2,918.81
31st month	-$2,918.81	-$14.59	-$2,933.41	-$3,743.41
32nd month	-$3,743.41	-$18.72	-$3,762.12	-$4,572.12
33rd month	-$4,572.12	-$22.86	-$4,594.98	-$5,404.98
34th month	-$5,404.98	-$27.02	-$5,432.01	-$6,242.01
35th month	-$6,242.01	-$31.21	-$6,273.22	-$7,083.22
36th month	-$7,083.22	-$35.42	-$7,118.63	-$7,928.63

a Look at the values towards the end of the spreadsheet. When the amount owing after the repayment becomes negative the loan has been repaid. The first negative, amount owing is in cell E35, which corresponds to the 27th payment month. Jordan will make 26 full payments and then a partial payment.

Jordan will repay the loan in the 27th payment month, but the last payment will be smaller than $810. It will be only $340.44.

b Add all the values in the interest column up to and including the interest for the 27th month. Entering a formula into the spreadsheet will be the easiest way.

The formula is =SUM(C9:C35).

Jordan will pay $1400.44 interest.

Exercise 4.04 Loan spreadsheets

Use the Reducible loans spreadsheet from NelsonNet to complete this exercise.

Reducible loans

Example
6

1 Samantha borrowed $5000 at 7% p.a. reducible interest to buy a motor scooter. Her monthly repayments are $350.

iStock.com/Bm

a How much interest will Samantha pay in the first month?

b How much of Samantha's second repayment will be interest?

c How much will Samantha owe immediately after she has made her 12th repayment?

d How long will it take Samantha to repay the loan?

e What is the value of Samantha's last repayment?

f How much interest will Samantha pay on the loan?

g Calculate the total amount Samantha will repay.

2 Nazim borrowed $6000 at 4.25% reducible interest to buy some new computer equipment. Each month Nazim repays $360.

a How long will it take Nazim to repay the loan?

b How much interest will Nazim pay?

c How much more than he borrowed will Nazim repay?

3 Billy's grandmother is going to lend him $7000 interest-free to buy his first car, provided that he repays her $400 each month.

a How long will it take Billy to repay the interest-free loan?

b If Billy had to borrow the money from the bank, he would be charged 9% p.a. reducible interest. How much interest is Billy saving with his grandmother's interest-free loan?

4 Arun wants to borrow $9000. The bank has offered him a reducible interest loan at 7.4% p.a. and suggested that he repay $400 each month.

a How long will it take Arun to repay the loan with monthly repayments of $400?

b Arun thinks he can afford to repay $550 per month. How long will it take him to repay the loan at $550 per month?

c Will Arun save any money by paying the loan off more quickly? Justify your answer.

d What general conclusion can you make about the advantage of repaying a reducible loan quickly?

e Is the same conclusion true for flat-rate loans? Explain your answer.

5 Lily made a spreadsheet of her reducing balance loan.

	A	B	C	D	E	F
1	**Reducible loans**					
2	Only enter data in cells shaded blue. Enter the loan and repayment amounts without any spaces, commas or $ sign.					
3						
4	Loan amount	$4,000.00				
5	Interest rate as a decimal	0.06				
6	Monthly repayment	$120.00				
7						
8		Amount owing at the beginning of the month	Interest charge for the month	Amount owing plus interest	Amount owing after the repayment	
9	1st month	$4,000.00	$20.00	$4,020.00	$3,900.00	
10	2nd month	$3,900.00	$19.50	$3,919.50	$3,799.50	
11	3rd month	$3,799.50	$19.00	$3,818.50	$3,698.50	
12	4th month	$3,698.50	$18.49	$3,716.99	$3,596.99	
13	5th month	$3,596.99	$17.98	$3,614.97	$3,494.97	
14	6th month	$3,494.97	$17.47	$3,512.45	$3,392.45	
15	7th month	$3,392.45	$16.96	$3,409.41	$3,289.41	
16	8th month	$3,289.41	$16.45	$3,305.86	$3,185.86	
17	9th month	$3,185.86	$15.93	$3,201.79	$3,081.79	
18	10th month	$3,081.79	$15.41	$3,097.20	$2,977.20	
19	11th month	$2,977.20	$14.89	$2,992.08	$2,872.08	
20	12th month	$2,872.08	$14.36	$2,886.44	$2,766.44	
21	13th month	$2,766.44	$13.83	$2,780.28	$2,660.28	
22	14th month	$2,660.28	$13.30	$2,673.58	$2,553.58	
23	15th month	$2,553.58	$12.77	$2,566.35	$2,446.35	
24	16th month	$2,446.35	$12.23	$2,458.58	$2,338.58	
25	17th month	$2,338.58	$11.69	$2,350.27	$2,230.27	
26	18th month	$2,230.27	$11.15	$2,241.42	$2,121.42	
27	19th month	$2,121.42	$10.61	$2,132.03	$2,012.03	
28	20th month	$2,012.03	$10.06	$2,022.09	$1,902.09	
29	21st month	$1,902.09	$9.51	$1,911.60	$1,791.60	
30	22nd month	$1,791.60	$8.96	$1,800.56	$1,680.56	
31	23rd month	$1,680.56	$8.40	$1,688.96	$1,568.96	
32	24th month	$1,568.96	$7.84	$1,576.80	$1,456.80	
33	25th month	$1,456.80	$7.28	$1,464.09	$1,344.09	
34	26th month	$1,344.09	$6.72	$1,350.81	$1,230.81	
35	27th month	$1,230.81	$6.15	$1,236.96	$1,116.96	
36	28th month	$1,116.96	$5.58	$1,122.55	$1,002.55	
37	29th month	$1,002.55	$5.01	$1,007.56	$887.56	
38	30th month	$887.56	$4.44	$892.00	$772.00	
39	31st month	$772.00	$3.86	$775.86	$655.86	
40	32nd month	$655.86	$3.28	$659.14	$539.14	
41	33rd month	$539.14	$2.70	$541.83	$421.83	
42	34th month	$421.83	$2.11	$423.94	$303.94	
43	35th month	$303.94	$1.52	$305.46	$185.46	
44	36th month	$185.46	$0.93	$186.39	$66.39	

What formulas did Lily enter in cells B9, C9, D9, E9 and B10?

6 Gabriel borrowed $30 000 from the bank to buy a car. The bank is charging him 7.5% p.a. monthly reducible finance and each month Gabriel repays $950.

Use a spreadsheet to determine the number of $950 repayments Gabriel will make and the total amount of interest he will pay on the loan.

7 Monique borrowed $6400 at 5.8% monthly reducible interest to go on an European holiday. Her monthly repayments are $520.

   a How long will it take Monique to repay the loan?

   b How much interest will Monique pay?

INVESTIGATION

WHICH LOAN?

25% DEPOSIT REQUIRED ON TERMS.

Term options: 8% flat rate interest for 3 years or reducible interest at 10% p.a. with monthly repayments of $544.

$16 000 cash

iStock.com/Rawpixel

Suzie wants to buy this car, but she will need to buy on terms because she doesn't have enough money to pay cash.

She thinks she should take the loan with the smaller interest rate, but she's heard that reducible interest can be better. What should she do?

> When you are borrowing money, the better option is the one with the smaller interest amount!

What you have to do

1 Make appropriate calculations to determine which of the two options is better.

2 Write an email to Suzie giving your advice about the better loan. Explain the reasons for your recommendation.

4.05 Online loan calculators

MoneySmart

Most people use online calculators to help them investigate and manage reducing balance loans. Find an online calculator to use in this section, for example, the mortgage calculator on the **MoneySmart** website.

EXAMPLE 7

Tom borrows $350 000 to buy an apartment in which to live. He will repay the loan in equal monthly instalments over 25 years at 7.5% p.a. monthly reducible interest. In addition, he will be charged a $10 monthly account-keeping fee.

a   How much are Tom's monthly repayments?

b   How much will Tom pay in interest and fees?

Solution

a   Enter the values into the home loan mortgage calculator.

Tom's monthly repayments will be $2596.

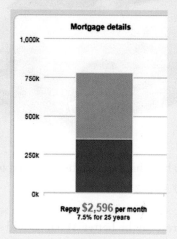

ASIC's MoneySmart website Mortage Calculator (https://www.moneysmart.gov.au/tools-and-resources/calculators-and-tools/mortgage-calculator). Date accessed March 2017

b   Interest and fees
= Total repaid − amount borrowed

We calculate the total amount Tom repays by multiplying the monthly repayments by the number of months in 25 years.

Write your answer.

Number of repayments = 25 × 12
                                              = 300

Total repaid = $2596 × 300
                     = $778 800

Interest and fees = $778 800 − $350 000
                             = $428 800

Tom's monthly repayments will be $2596 and he will pay $428 800 in interest and fees.

EXAMPLE 8

MoneySmart

Robby overspent on his credit card and bought $5000 worth of items with it. His credit card charges 21% p.a. monthly reducible interest, but he can afford to repay only $120 per month.

a How long will it take Robby to repay the $5000 credit card bill if he makes $120 repayments each month?

b How much will it cost Robby to pay off his $5000 credit card bill?

Solution

a Enter the values into the personal loan calculator. Use the calculator in the tab 'How can I repay my loan sooner?' Ensure fees are set to $0.

The online calculator shows that it will take Robby 6 years and 4 months to pay off the loan.

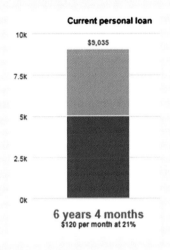

Current personal loan

Amount owing:

$5,000

Repayment:

$120   Monthly ⌄

Interest rate:

21.00%

Fees:

$0   Monthly ⌄

ASIC's MoneySmart website Mortgage Calculator (https://www.moneysmart.gov.au/tools-and-resources/calculators-and-tools/personal-loan-calculator). Date accessed March 2017

b The total amount Robby has to pay is $120 multiplied by the number of repayments.

$$\text{Number of repayments} = 6 \times 12 + 4$$
$$= 76$$

$$\text{Total Robby will pay} = 76 \times \$120$$
$$= \$9120$$

Write your answer.

Robby will pay $9120 in order to pay off his $5000 credit card bill.

MoneySmart

Exercise 4.05 Online loan calculators

You will need access to online calculators to complete this set of exercises.
Use the personal loan calculator to answer Questions **1** to **3**.

Example 7

1 Claire is going to borrow $15 000 at 9.8% p.a. monthly reducible interest to start a business. She plans to repay the loan in monthly repayments over 5 years.

   a How much are her monthly repayments?

   b Calculate the total amount Claire will pay.

   c How much interest will Claire pay during the 5 years?

2 Mark wants to borrow $40 000 to buy an SUV. He can get the money from a finance company at 17% p.a. reducible interest.

   a Calculate the monthly repayments if Mark takes the loan over 15 years.

   b How much interest will Mark pay if he takes the loan over 15 years?

   c Give 2 reasons why 15 years isn't a suitable term for a car loan.

   d If you were Mark, what would you do?

Example 8

3 Ewan wants to borrow $10 000 to go on a trip to Canada. His bank will lend him the money at 8.75% p.a. reducible finance with no monthly account fees. Ewan can afford to repay $650 per month.

   a Use the 'How can I repay my loan sooner' tab on the personal loan calculator to calculate the number of $650 repayments Ewan will make to repay the loan.

   b How much interest will Ewan pay?

4 When Hoa bought her home she borrowed $250 000 at 8% p.a. monthly reducing interest over 30 years and her monthly repayments were $1834.40. The graph shows the balance B, the amount she owed on her loan after she made N repayments. Use the graph to answer the following questions.

Repaying a $250 000 loan over 30 years

Graph: Balance owing (B) vs Number of repayments made (N)

- 20th payment interest = $1644
- 40th payment interest = $1617
- (90, $229 405)
- $N = 257$ interest = $915.20
- (269, $124 850)
- $250 000 starting balance
- N-axis marks at 90, 180, 270, 360

   a Of the 20th payment, $1644 was interest. How much of the repayment came off the balance?

   b Why is the amount of interest in the 40th repayment less than the interest in the 20th repayment?

c Calculate the total amount Hoa repaid during the first 90 repayments.

d Find the amount Hoa still owed on the loan after she made 90 repayments.

e Of the 257th repayment, $915.20 was interest. How much of the 257th repayment came off the loan? Explain the significance of this amount.

f How long did it take for the balance of the loan to be less than half of the amount Hoa borrowed?

g How long did it take Hoa to repay the second half of the loan?

h When we borrow money at reducible interest, when do we pay the most interest?

5 Karl borrowed $500 000 at 7.9% p.a. monthly reducible interest to buy a house and made monthly repayments of $5000. After 18 months, he decided that he couldn't afford the repayments and he sold the house.

Karl said: "I've paid 18 lots of $5000, 18 × $5000 is $90 000. Subtract a bit for interest. I'll have repaid about $75 000."

Explain the error in Karl's thinking.

6 When Lara borrowed $200 000 to buy an apartment, the bank told her the repayments would be $830 per month. Lara decided that she could afford to repay $930 monthly. The graph shows the balance of a $200 000 loan with monthly repayments of $830 and $930.

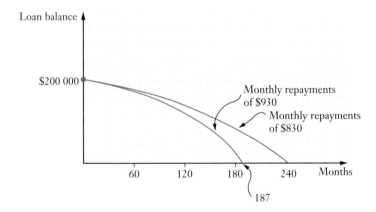

a How long will Lara take to repay the loan if she makes monthly repayments of $830 and $930?

b How much will Lara save by making monthly repayments of $930?

c Why do you think some people may choose to take the smaller monthly repayments even though they will have to pay more in the long run?

Use the mortgage calculator to answer the remaining questions in this exercise.

7 Michael borrowed $260 000 at 6.75% p.a. over 15 years.

  a How much were his monthly repayments?

  b Interest rates fell to 6.5% p.a. How much cheaper were Michael's monthly repayments after the fall in interest rates?

8 When interest rates go up, loan repayments increase. When Jason borrowed $170 000, the interest rate was 6.75% p.a. and he took the loan over 10 years.

  a How much was Jason's monthly repayment when he took out the loan?

  b By how much did Jason's monthly repayments increase when the interest rate went up to 7% p.a.?

  c How much more did Jason have to repay each year at 7% p.a. compared to the original annual amount?

9 Zhi is buying a house. He is going to borrow $240 000 at 8% p.a. He is trying to decide whether to take the loan over 20 or 30 years.

  a How much more are the monthly repayments over 20 years than over 30 years?

  b How much more interest will he pay if he takes the loan over 30 years than over 20 years?

  c If you were Zhi, would you take the loan over 20 or 30 years? Give reasons.

10 During the 1980s, interest rates rose as high as 18% p.a. How much more were the monthly repayments on a $150 000 loan over 20 years in the 1980s compared to the same loan at 7% p.a. in the early 2010s?

11 Sarah is deciding between the 2 home loans shown in the table.
She is borrowing $140 000 over 25 years.

	Interest rate	Loan establishment fee	Annual loan fee	Mortgage discharge fee
'Big 4 bank'	7.1% p.a.	$320	$248	$228
Small mortgage company	7.0% p.a.	$598	$76	$314

  a Calculate the value of the monthly repayments for each loan.

  b Copy and complete this table to help Sarah determine the better loan.

	Big 4 bank	Small mortgage company
Loan establishment fee		
Mortgage discharge fee		
Total annual loan fee over 25 years		
Total monthly repayments		
Total cost of the loan		

  c Which 2 features of a loan – the interest rate, loan establishment fee, annual loan fee or mortgage discharge fee – most influence the total cost of a loan?

  d Which of the 2 loans do you recommend Sarah take? Why?

MoneySmart

INVESTIGATION

MAKING SMART REPAYMENTS

Let's examine the effects of making home loan repayments more frequently and increasing the size of each repayment. Visit the **MoneySmart** website and search for **Mortgage calculators**. Select **How can I repay my home loan sooner?**

1 Declan has a $250 000 loan at 8% p.a. and he can afford to repay $3000 per month or $1500 per fortnight. Does paying off a home loan fortnightly instead of monthly make any difference?

Use the online calculator to help you copy and complete this table.

	Monthly repayments	Fortnightly repayments
Value of the repayment	$3000	$1500
Time to pay off the loan		
Total repayments		
Total interest		

2 Repeat Question **1** for a $300 000 loan at 12% p.a. with a monthly repayment of $5000 and a fortnightly repayment of $2500.

3 Investigate some other loan amounts (principals), but make sure the monthly repayment is twice the value of the fortnightly repayment.

4 What conclusions can you make? Does repaying half the monthly repayment each fortnight make any difference to the loan?

5 Madison borrowed $250 000 at 8% p.a. with a monthly repayment of $2000. Select the tab **How can I repay my home loan sooner**? Use the calculator to find the term (length of time) of Madison's loan. Record the total amount and the interest paid by Madison.

6 Does increasing the monthly repayments by $20 make any difference? Change the repayment from $2000 to $2020. Record the new term of the loan, as well as the new total amount and interest paid.

7 Repeat Question **6** for a larger monthly repayment.

8 What conclusions can you make? Does repaying a larger amount make much difference to the loan?

4.06 Credit cards

Credit card
statement

Investigating
credit cards

Using a credit card is a convenient way to shop and pay bills but if used recklessly, a big financial problem can result. Credit card interest is calculated daily and sometimes the rates are more than 20% p.a.

Credit card statements contain a lot of information. Often, the most important information is in small print.

EXAMPLE 9

(a) East Coast Bank

Loyalty Card

Account Number		
4564 7170 2243 6120		
Due Date		
10 Oct 18	**(g)**	
Closing Balance		
121.39	**(e)**	
Minimum Payment Due		
10.00	**(f)**	

‖ıllıılı·ı·ıllılıllıllıırı·lıı·ıllllıll·······ıll·lı·ıllıı·lı 021

(b) Ms Jane Smith
75 Valley Way
WYONG NSW 2259

Amount Paid (details on the reverse)

$ _____

- ✂
(Cut along this dotted line)

Cardholder Name	Account Number	Customer Number	Credit Limit	Available Credit
Ms Jane Smith	4564 7170 2243 6120	04389804	27,500	27,378.61

No. of Days in Statement Cycle	Statement From	Statement To	Minimum Payment Due	Due Date	Opening Balance
(d) 30	27 Aug 18	25 Sep 18	10.00	10 Oct 18	1,453.17-

Summary of Changes in Your Account Since Last Statement

From Your Opening Balance of	We Deducted Payments And Other Credits	And We Added		To Arrive at Your Closing Balance of	Past Due / Overlimit is	Your Minimum Payment Including Past Due / Overlimit is
		New Purchases Cash Advances And Other Debits	Fees, Government & Interest Charges			
1,453.17-	1,110.00-	2,684.56	0.00	121.39	0.00	10.00

Interest rates:

Cash advances	21.6% p.a. **(c)**
Purchases	19.99% p.a.

Minimum Repayment Warning: If you make only the minimum payment each month, you will pay more interest and it will take you longer to pay off your balance. For example:

If you make no additional charges using this card and each month you pay...	You will pay off the Closing Balance shown on this statement in about...	And you will end up paying estimated total interest charges of...
Only the minimum payment	1 year 2 months	$15.30

Having trouble making repayments? If you are having difficulty making credit card repayments, please contact us.
Remember: Pay the minimum payment to avoid the $9 missed payment charge. **(h)**

ISBN 9780170413534

Use the credit card statement to answer the following questions.

a In what financial institution is the credit card held?

b What is the name and address of the card holder?

c What interest rates apply to the account?

d How many days are in the statement cycle?

e How much does the account holder owe on the card?

f What is the minimum repayment?

g On what date is the repayment due?

h What happens if you don't pay the minimum repayment by the due date?

Solution

See labels **a** to **h** on the credit card statement.

a	Top left of the statement	East Coast Bank
b	Top left	Jane Smith, 75 Valley Way, Wyong 2259, NSW
c	Bottom left	The bank charges 21.6% p.a. interest for cash advances and 19.99% p.a. interest for purchases.
d	Middle left	The bill is for 30 days.
e	Closing balance, top right	The account holder owes $121.39.
f	Top right	The minimum repayment is $10.
g	Top right	The date due is 10th October 2018.
h	Bottom	The bank will charge a $9 fee and daily interest until the total amount is repaid.

Exercise 4.06 Credit cards

1

East Coast Bank
Loyalty Card

Account Number
4564 7170 2243 6120

Due Date
09 Nov 18

Closing Balance
16.29

Minimum Payment Due
10.00

Amount Paid (details on the reverse)

$ [ ]

Ilμ‖‖‖·‖·‖‖‖·‖‖‖‖‖‖‖‖‖‖···········‖‖‖·‖‖‖μ‖·   021

Ms Jane Smith
75 Valley Way
WYONG NSW 2259

- ✂
(Cut along this dotted line)

Cardholder Name	Account Number	Customer Number	Credit Limit	Available Credit
Ms Jane Smith	4564 7170 2243 6120	04389804	27,500	27,483.71

No. of Days in Statement Cycle	Statement From	Statement To	Minimum Payment Due	Due Date	Opening Balance
30	26 Sep 18	25 Oct 18	0.00	09 Nov 18	133.97

Summary of Changes in Your Account Since Last Statement

From Your Opening Balance of	We Deducted Payments And Other Credits	And We Added		To Arrive at Your Closing Balance of	Past Due / Overlimit is	Your Minimum Payment Including Past Due / Overlimit is
		New Purchases Cash Advances And Other Debits	Fees, Government & Interest Charges			
133.97	2,280.00-	2,162.32	0.00	16.29	0.00	10.00

Interest rates:

Cash advances	21.6% p.a.
Purchases	19.99% p.a.

Minimum Repayment Warning: If you make only the minimum payment each month, you will pay more interest and it will take you longer to pay off your balance. For example:

If you make no additional charges using this card and each month you pay...	You will pay off the Closing Balance shown on this statement in about...	And you will end up paying estimated total interest charges of...
Only the minimum payment	2 months	$1.26

Having trouble making repayments? If you are having difficulty making credit card repayments, please contact us.
Remember: Pay the minimum payment to avoid the $9 missed payment charge.

---

a How much does Jane owe on her credit card?

b When does she have to make the repayment?

c What fee will the bank charge Jane if she doesn't repay the money she owes by the due date?

d Jane only uses her credit card to make purchases. What interest rate will she be charged if she doesn't repay the amount owing?

e What is Jane's customer number?

f How much is Jane's credit limit on this credit card?

ISBN 9780170413534

2 Here is Seb's credit card statement.

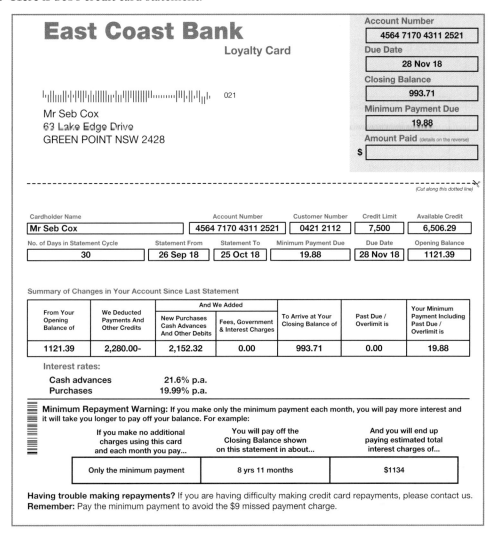

East Coast Bank
Loyalty Card

Account Number	
4564 7170 4311 2521	
Due Date	
28 Nov 18	
Closing Balance	
993.71	
Minimum Payment Due	
19.88	
Amount Paid (details on the reverse)	
$	

||||l|l·l·|ll|l|ll|ll|l|lr·lll||llll|ll·······||l|·l|l|l| 021

Mr Seb Cox
63 Lake Edge Drive
GREEN POINT NSW 2428

(Cut along this dotted line)

Cardholder Name	Account Number	Customer Number	Credit Limit	Available Credit
Mr Seb Cox	4564 7170 4311 2521	0421 2112	7,500	6,506.29

No. of Days in Statement Cycle	Statement From	Statement To	Minimum Payment Due	Due Date	Opening Balance
30	26 Sep 18	25 Oct 18	19.88	28 Nov 18	1121.39

Summary of Changes in Your Account Since Last Statement

From Your Opening Balance of	We Deducted Payments And Other Credits	And We Added		To Arrive at Your Closing Balance of	Past Due / Overlimit is	Your Minimum Payment Including Past Due / Overlimit is
		New Purchases Cash Advances And Other Debits	Fees, Government & Interest Charges			
1121.39	2,280.00-	2,152.32	0.00	993.71	0.00	19.88

Interest rates:

Cash advances   21.6% p.a.
Purchases   19.99% p.a.

Minimum Repayment Warning: If you make only the minimum payment each month, you will pay more interest and it will take you longer to pay off your balance. For example:

If you make no additional charges using this card and each month you pay...	You will pay off the Closing Balance shown on this statement in about...	And you will end up paying estimated total interest charges of...
Only the minimum payment	8 yrs 11 months	$1134

Having trouble making repayments? If you are having difficulty making credit card repayments, please contact us.
Remember: Pay the minimum payment to avoid the $9 missed payment charge.

a How much does Seb owe on his credit card?

b How much did Seb owe at the beginning of the statement period?

c How much did Seb repay during the statement cycle?

d When does Seb need to make a repayment and what is the minimum repayment?

e If Seb doesn't spend any more on his credit card and he makes only the minimum monthly repayments:

  i how long will it take him to repay the credit card debt?

  ii how much interest will he pay?

  iii Calculate the total amount Seb will repay the East Coast Bank to pay off his credit card debt.

f Find the value of the purchases Seb made on his credit card during the billing cycle.

3 The second page of Inge's credit card statement shows the transactions she made on her card.

East Coast Bank

Loyalty Card

PAGE 2

Loyalty Points Balance Summary	
Opening Balance	28,133.00
Points Earned	2,684.00
Points Redeemed	0.00
Closing Balance	30,817.00
Points Status	Available

Date of Transaction	Description			Debits	Credits (–)
23 Aug 18	SUN PRINCESS SDM	LONDON	GB	147.70	
26 Aug 18	PACIFIC SMILES GROUP L	TUGGERAH	AU	45.00	
27 Aug 18	THE ATHLETES FOOT	TUGGERAH	AU	219.95	
27 Aug 18	CALTEX BERKELY VALE	BERKELEY VALE	AU	60.52	
27 Aug 18	LOWES MANHATTAN	TUGGERAH	AU	29.95	
29 Aug 18	AVG	Sandringham	AU	106.79	
29 Aug 18	SUN PRINCESS	ST LEONARDS	AU	775.39	
30 Aug 18	PACIFIC SMILES GROUP L	TUGGERAH	AU	107.00	
31 Aug 18	INDULGE HAIR DESIGN	TUMBI UMBI	AU	118.00	
03 Sep 18	PAYMENT - CASH THANK YOU				40.00
10 Sep 18	PAYMENT - CASH THANK YOU				120.00
11 Sep 18	CALTEX BERKELEY VALE	BERKELEY VALE	AU	57.64	
15 Sep 18	HARRIS SCARFE AUSTRA	TUGGERAH	AU	100.00	
15 Sep 18	ADAIRS - TUGGERAH	TUGGERAH	AU	301.00	
18 Sep 18	PAYMENT - CASH THANK YOU				460.00
19 Sep 18	DR PETER SOLOMON	WOY WOY	AU	270.00	
21 Sep 18	CALTEX WOOLWORTHS LO	LONG JETTY	AU	46.67	
22 Sep 18	TUMBI TYRES & AUTO	TUMBI UMBI	AU	91.00	
22 Sep 18	ADAIRS - TUGGERAH	TUGGERAH	AU	37.95	
24 Sep 18	PAYMENT - CASH THANK YOU				270.00
25 Sep 18	PAYMENT - CASH THANK YOU				220.00
25 Sep 18	FINCORA PTY LTD	GOSFORD	AU	170.00	

Transaction Type	Annual Percentage Rate	Expiry	Closing Balance
Cash Advances	21.59		0.00
Purchases	19.99		421.50

a On 27 August, Inge bought a new pair of gym shoes. How much did the gym shoes cost and at what shop did she buy them?

b Inge visited the doctor on 19 September. What was the doctor's name and how much did she have to pay?

c During the billing cycle, Inge bought petrol from three Caltex service stations. How much did Inge charge to her credit card for petrol?

d During August, Inge went on a cruise on the Sun Princess ship. How much did the Sun Princess charge to Inge's credit card from 23 August to 30 August?

e Inge had streaks put in her hair. Which hairdresser did she go to and how much did the streaks cost?

f During September, Inge deposited five payments into her credit card. Calculate the total amount Inge deposited.

g How many loyalty points does Inge have?

h Inge uses her loyalty points to buy gift cards for her friends. A $50 gift card costs 14 500 points. How many $50 gift cards can Inge buy with her points and how many points will be left?

i There is an annual $75 fee on Inge's credit card that she can pay with 'Loyalty points'. Based on the points required to buy a $50 gift card, estimate the number of points Inge needs to pay for her annual card fee.

Managing my
credit card

INVESTIGATION

HOW CAN I KEEP TRACK OF MY CREDIT CARD?

Using a spreadsheet is a great way to keep track of your finances. Ask your teacher to download the 'Managing my credit card' spreadsheet from NelsonNet.

Imagine you have paid for the following items with your credit card.

Date	Amount	Description
2 April	$68	Bought Easter eggs for the Easter egg hunt
8 April	$52	Bought petrol
12 April	$130	Deposit from my part-time job
18 April	$85	Monthly gym fee
19 April	$22	Hair cut
24 April	$210	Car repairs
29 April	$140	Deposit from my day job

1 Update the spreadsheet and determine whether you owe money or whether your credit card is in credit.

2 Record the size of any credit or amount owing on the account.

3 Imagine you pay all your bills with a credit card. Complete the spreadsheet for your monthly expenses and determine the amount you need to budget to pay off your credit card each month.

WHAT IF I CAN'T AFFORD TO PAY MY CREDIT CARD BILL?

Tom spent $1800 on presents for his family and friends and he charged the entire amount to his credit card. Now he can't afford to repay $1800 and he is being charged 20% p.a. interest. What can Tom do?

MoneySmart

1 Visit the **MoneySmart** website and search for **Credit card calculator**.

2 Enter $1800 for the card balance and 20% for the interest rate.

3 Record the amount Tom will have to repay and the length of time it will take him to pay off this credit card debt if he makes only the minimum monthly repayments.

4 Select **Higher repayments** and record the time it will take Tom to pay off the loan, as well as the total amount it will cost him if he makes monthly repayments of $90.

5 Increase Tom's monthly repayments to $120. What affect will this change have?

6 Sometimes, banks have special low-interest credit cards to attract new customers. What will be the result if Tom transfers his debt to a different bank that is offering 2% p.a. interest for new customers and he can make monthly repayments of $150?

7 Write a short letter of advice to Tom.

WHY ARE CREDIT CARD INTEREST RATES SO HIGH?

1 Search online for the video 'High credit card interest rates'.

2 View the video and explain why credit card interest rates are so high.

3 Also, view the video 'Credit card debt explained with a glass of water' to understand why large amounts of interest can be involved in paying off credit card debt.

Credit card
calculations

4.07 Credit card interest

Credit cards are very popular because they are convenient and they make online purchases easy. However, wise credit card users are aware of the fees and interest that financial institutions can charge. The interest charged can be compound or flat. Some credit cards have an **interest-free period** (for example, 55 days), during which you will not be charged interest if you repay the total amount owing. Otherwise, interest is charged *from the date of purchase*.

EXAMPLE 10

Amber's credit card charges 20% p.a. compound interest on all purchases. The interest is calculated daily. How much interest was she charged for purchasing $1900 in airfares on her card for 16 days?

Solution

$FV = PV(1 + r)^n$ where $PV = \$1900$, $r = 0.2 \div 365$, $n = 16$. Because the time is in days, the interest rate must be in days too.

> Remember! 1 year = 365 days.

$FV = \$1900(1 + 0.20 \div 365)^{16}$

$= \$1916.7261 \ldots$

Interest charged $= \$1916.7261 \ldots - \$1900$

$= \$16.7261 \ldots$

$\approx \$16.73$

EXAMPLE 11

Tim owes $1500 on his credit card. The flat interest rate is 20% p.a. and the minimum monthly payment is 2%.

a Calculate the interest Tim will be charged for 1 month.

b What is the minimum monthly payment?

c Calculate the amount Tim will owe after 1 month if he pays only the minimum monthly payment.

d By how much will the balance of Tim's credit card bill reduce after 1 month if he makes the minimum payment only?

Solution

a $I = Prn$, $n = 1$ month, so the interest rate has to be monthly.

$P = \$1500$, $r = 0.2 \div 12$, $n = 1$.

Interest $= 1500 \times (0.2 \div 12) \times 1$
$\quad\quad\quad = \$25$
Tim will be charged $25 interest.

b Minimum monthly payment is 2%.

Minimum payment $= 2\% \times 1500$
$\quad\quad\quad\quad\quad\quad = \$30$

c Amount owing
$= \$1500 + \text{interest} - \text{repayment}$

Amount owing $= \$1500 + \$25 - \$30$
$\quad\quad\quad\quad\quad = \$1495$

d Original balance − final balance.

Decrease in balance $= \$1500 - \$1495$
$\quad\quad\quad\quad\quad\quad\quad = \$5$
The balance of Tim's account will go down by only $5 even though he repaid $30.

Exercise 4.07 Credit card interest

1 Luka paid the $920 balance on his credit card after 25 days. His card charges 20% p.a. interest compounded daily. How much interest was he charged?

Example 10

2 Olive's credit card charges 19.5% p.a. interest, compounded daily. She bought furniture for $2650 on her card and paid in 30 days. How much interest was she charged?

3 Patrick's credit card charges him 0.056% daily compound interest on purchases.

a What annual rate of interest is equivalent to a daily rate of 0.056%?

b Explain why 0.056% = 0.00056 as a decimal.

c Patrick owed $2400 on his credit card for 42 days. How much interest was he charged?

4 When Saskia didn't pay her $760 credit card bill on time, she was charged a $9 bank fee and compound interest at 0.056% per day. She repaid the $760 she owed plus the fee and interest for 7 days. How much did she have to pay?

Example
11

5 Emily owes $3600 on her credit card and she can only afford to repay the minimum amount, which is 2% of the amount she owes. The flat interest rate on her credit card is 19.75% p.a.

 a How much interest will Emily be charged for 1 month?

 b Calculate the minimum repayment.

 c How much will Emily still owe on her credit card after she makes the minimum repayment?

6 Holly's Palm Tree Cove credit card has an interest-free period.

 • No interest charged if the account is paid in full on or before the due date.

 • If the account is paid after the due date, a 1.5% late fee is charged on the balance plus 0.04953% daily flat-rate interest.

Palm Tree Cove		Credit Card		
Account Number PTC 234234		Statement date 13 Apr 19		
Account Name	Miss Holly Citizen			
Date	Transaction Details	Debits	Credits	Balance
14 Mar	Opening Balance			$636.70
16 Mar	Caltex — Ryde	$182.50		$819.20
17 Mar	Payment/Thank you		$637.00	$182.20
24 Mar	Florist — Penrith	$45.00		$227.20
3 Apr	Palm Tree Superstore	$251.00		$478.20
13 Apr	Government Charges	$0.18		$478.38
			Closing Balance	$478.38
	Credit Limit $2500	Available Credit	$2021.62	
	Credit Charges 18% p.a. 0.04953% daily			
	Overdue amount $0.00	Minimum due $15.00	Due Date 27 Apr 19	

 a How much is Holly's credit limit?

 b How much credit is available now? Explain why this amount is not the same as her credit limit.

 c What is the minimum amount Holly can repay this month?

 d How much does Holly need to repay to avoid any fees or interest?

 e By what date should Holly pay the account?

 f Holly was slow paying the account. She paid the account 2 days late.

  i What late fee was she charged?

  ii How much flat-rate interest will she have to pay?

4.08 It's your money and your life

You are 22 years old and you are enjoying life. You work $5\frac{1}{2}$ days per week and like to relax and have fun whenever you can. You haven't worried too much about your finances in the past but recently you're finding it hard to pay your bills. As you work through the questions in Exercise 4.08, you'll learn about your money concerns.

Exercise 4.08 It's your money and your life

Your income and expenses

You have completed your apprenticeship and have been working full-time for 3 years. You have just moved out of home and are renting a unit. Your take-home pay is $820 per week and your expenses are:

- Fortnightly rent $776
- Car repayments $93 per week
- Food and clothing $165 per week

1 What percentage of your take-home pay goes to rent?

2 How much of your take-home pay is left each week after you pay rent, car payments, food and clothes?

3 When you moved into your unit you wanted some nice furniture and entertainment equipment. You bought an $8000 'new home package' from a discount store. The store gave you a special 'no interest and no repayments for 24 months' deal. You didn't read the small print carefully.

NEW HOME PACKAGE only $8000
No interest, no repayments for 24 months!

The entire purchase price is to be repaid 24 months from the contract date. If the entire payment isn't received by the due date, then an interest charge of 29.45% will be added to the amount owing. Continuing finance will be subject to interest at 29.45% p.a. calculated on the amount owing at the end of each month.

a Initially, you deposited $75 each week into an account to save the money to pay for the new home package in 24 months. If you continue to make these savings, will you have enough money to pay the bill in 24 months? Justify your answer.

b You want to spend this money on other things and have stopped saving. What penalty is the discount store going to charge when you don't pay for the new home package by the due date?

c You haven't repaid the money on the due date and instead start to repay $400 per month. How much will you owe after 1 month?

d If you don't repay the entire loan on the due date, the loan will operate like a credit card. Use a credit card simulator, with interest at 29.45% p.a., to determine the total amount you will have to repay and the time it will take you. There is a simulator on the **MoneySmart** website.

MoneySmart

4 Often, you don't have the cash to pay for the things you want to buy, so you charge them to your credit card. At present, you owe $3210 on your credit card and your credit card company charges 22% p.a. interest on overdue amounts.

a Use a credit card simulator to determine the minimum monthly amount the credit card company requires you to pay.

b How long will it take you to repay your credit card debt if you pay only the minimum amount?

c How much will you pay in interest if you make monthly repayments of $164 until the loan is repaid?

d You saw a TV ad about another bank offering interest at only 6% p.a. on balances transferred from other credit cards. If you transfer your credit card balance to this bank and make monthly repayments of $164, how much interest will you save compared to your current bank which is charging 22% p.a. interest?

5 You know you are in financial trouble when your electricity bill arrives. You have 3 weeks to find $490 and you don't have any money. You use your credit card to pay the bill, then start to panic about your financial problems. You see a Money Guru on TV and decide to write to the guru for advice.

Dear Money Guru

My credit card's maxed out and I can't afford the payments. When I make a payment most of it goes in interest and I still owe almost as much as before I made the payment. What can I do?

Signed,
Financially Drowning

Use this word list to copy and complete the Money Guru's reply.

credit card   difficulties   electricity   every day
huge   immediately   instalments   limited
low   minimum   owe   rush
stop   temporary

Dear Financially Drowning,

Don't panic! There are several strategies and options available to you. You haven't indicated whether your problem is ____a____ because you're short of money at the moment or whether it's a bigger problem but, either way, there are two things you should do ____b____.

Contact your bank now. Ask to speak with someone from the hardship department about managing your ____c____ debt. Explain that you are facing some financial ____d____ and ask them to vary the terms of your credit card agreement.

____e____ using your credit card for anything except an emergency situation.

Have you seen any of the current advertisements from other credit card companies? Several companies are offering very ____f____ interest rates for balances transferred to them when you change to their brand of credit card. The low interest rate only lasts for a ____g____ time, but it may help you get on top of the money you ____h____.

You haven't indicated whether you are having trouble paying your ____i____ or phone bills. If you are, don't pay them with your credit card. Contact the relevant company and arrange to pay the bills by ____j____.

Now, think about your lifestyle. Do you have an unnecessary expense you could delete? Do you spend money on a take-away coffee on the way to work? Your coffee might only cost $4, but if you buy one ____k____ it's costing you $80 or more a month. Stop the coffee and use the saving to pay off your debt. Remember, if you only pay the ____l____ amount each month, it will take a long time to pay off the debt and you'll pay a ____m____ amount of interest.

If you need more help, phone the financial counselling hotline on 1800 007 007. Be very wary of financial brokers who make unrealistic promises and try to ____n____ you into signing documents. They won't be acting in your best interests.

The Money Guru

---

6 You decide that you can't afford to rent a unit. You arrange to move into a small flat in your sister's garage. You are going to pay your sister rent of $100 per week and use the money you save from your unit rent to repay your debts. Which of your debts, the discount store debt or the credit card debt, should you concentrate on paying off first? Give a reason for your answer.

INVESTIGATION

CAN YOU AFFORD TO BUY A HOME?

Could you buy a home at age 25? Copy the table below or download it from NelsonNet.
Complete it by answering the questions on the next page.

Name					
1 Future job					
2 Annual gross pay					
3 Monthly gross pay					
4 Maximum monthly repayment					
5 Savings interest rate					
6 Deposit					
7 Mortgage interest rate					
		15 years	20 years	25 years	30 years
8 Maximum loan amount					
9 Monthly repayment					
10 Total repayments					
11 Total repayments if paid fortnightly					
12 Amount you can afford to spend on a property					
13 A property you can afford in a suitable location. Paste photo here.	Address				
	Price				
14 Net monthly income after deducting income tax and mortgage repayments.					
15 Five more expenses from my net monthly income.					
16 Strategies to make buying a property more affordable.					

MoneySmart

1 Write the type of occupation you plan to have when you are 25 years old; for example, panel beater, vet nurse, childcare worker.

2 Research the annual gross pay you will receive in this occupation. Do not include any overtime.

3 Calculate your monthly gross pay.

4 Calculate 30% of your monthly pay. This is the maximum amount your monthly repayment can be.

5 Research the current savings interest rate.

6 Suppose you are going to save 30% of your income each month for 5 years at the current savings interest rate. Use an appropriate calculator on the **MoneySmart** website to determine your savings at the end of 5 years. This amount is your deposit.

7 Research the current mortgage interest rate.

8 Use the **How much can I borrow?** mortgage calculator on the MoneySmart website to determine the maximum amount you can borrow over 15, 20, 25 and 30 years.

9 Record the monthly repayments in row 9 of the table.

10 Record the total you will repay in row 10 of the table.

11 Use the online calculator to determine the total you will repay if you make half the monthly repayment each fortnight. Record these amounts in row 11 of the table.

12 Calculate the amount you can afford to spend on a property by adding the amount you can borrow to the deposit you have saved.

13 Visit a real estate website and search for a property you can afford to buy in a suitable location. Paste a photograph of the property in the table. Record the address of the property and the selling price.

14 Assume that you will pay 25% of your gross pay in income tax. Subtract this tax and your monthly mortgage repayment from your gross pay. This amount represents the net pay you will have left each month. Write it in line 14 of the table.

15 List 5 different things you will have to pay out of the amount remaining in line 15.

16 Buying your first property is financially challenging. In a group, discuss strategies you could use to make it easier. Record the strategies in line 16 of the table.

COMPLETE THE BLANKS

Use the listed words below to copy and complete a summary of this chapter.

credit union	declining-balance	depreciation	down	flat
fortnightly	huge	interest	minimum	money
real estate	reducible	repay	salvage	same

The value of items we buy, such as cars, furniture, tools and clothes depreciate as they age. The older most items are, the less they are worth. **1**_____ is the value an item loses and its **2**_____ value is what it is worth as it ages. In straight-line depreciation, an item loses the same amount of value each year. With **3**_____ depreciation, the depreciation is a percentage of the previous year's value, so the amount the item depreciates by reduces as the item ages.

When we borrow money from a bank, building society or **4**_____ we have to pay interest. There are two kinds of **5**_____: flat-rate or simple interest and **6**_____ interest. Often, finance companies charge **7**_____ rate interest and banks charge reducible interest. When we borrow money at flat-rate interest, we pay the **8**_____ amount of interest every year, irrespective of the amount we actually owe, until we have repaid the loan. When we borrow under reducible interest, our interest payments go **9**_____ as we repay the money we owe.

When we borrow a lot of money over a long time; for example, when we buy **10** _____ _____, the total amount we repay can be **11**_____. Even small changes in interest rates can make a big difference to the amount we have to **12**_____. If we can arrange to repay more than the **13**_____ amount required each month, or repay half the monthly repayment **14**_____ we can save a lot of **15**_____ in interest.

SOLUTION TO THE CHAPTER PROBLEM

Problem

Leslie borrowed $240 000 to buy an apartment in which to live. She is going to repay the loan plus interest in monthly instalments at 7.8% p.a. monthly reducible finance.

If Leslie borrows the money over 15 years, the monthly repayments will be $2265.95 and the repayments for the loan over 30 years will be $1727.70.

a Calculate the total amount Leslie will repay if she takes the loan for 15 years or for 30 years.

b Suggest a reason why Leslie decided to take the loan for 30 years.

Solution

a Total amount repaid = value of the monthly repayment × number of years × 12

Loan over 15 years

Total amount repaid = $2265.95 × 15 × 12

= $407 871

Loan over 30 years

Total amount repaid = $1727.70 × 30 × 12

= $621 972

b Even though she will have to repay $214 101 more if she takes the loan over 30 years, Leslie could choose to do this because she can't afford the monthly repayments for the loan over 15 years. Alternative strategies that Leslie could consider include investigating whether she can afford the loan repayments over 20 or 25 years, or whether she can find a suitable, less expensive apartment.

- What part or parts of this chapter will be important in your life after school?

- List some items that people use in their work that depreciate.

- Do you know anyone in financial difficulty as a result of borrowing too much money to buy property or from spending more than they could afford on their credit card? Describe what happened.

- How do you plan to afford to pay for expensive items that you may need in the future?

- Are there any parts of this chapter that you don't understand? If yes, ask your teacher for help.

Copy and complete this mind map of the topic, adding detail to its branches and using pictures, symbols and colour where needed. Ask your teacher to check your work.

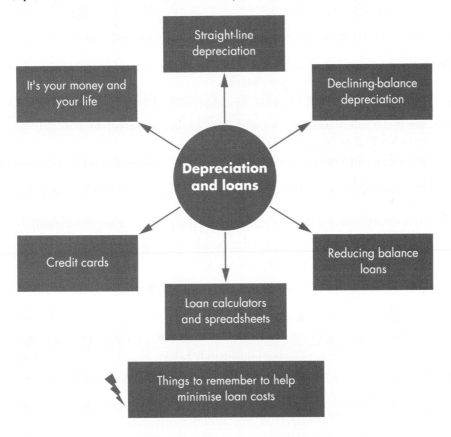

4. TEST YOURSELF

Exercise
4.01

1 Gillian bought a new chainsaw for $520. During the first year it depreciated by 10% of its original value and in the second year it depreciated by $40.

   a Calculate the chainsaw's salvage value when it was 2 years old.

   b How much did Gillian's chainsaw depreciate in the 2 years?

Exercises
4.01, 4.02

2 Every year, Bruce's golf clubs lose $50 in value. When the golf clubs were 1 year old, they were worth $860.

   a How much were Bruce's golf clubs worth when they were new?

   b Calculate the salvage value of Bruce's golf clubs when they are 4 years old.

Exercise
4.02

3 Nathan is using declining-balance depreciation at 22% p.a. to calculate the value of his earth-moving equipment. The equipment was valued at $125 000 when it was new.

   a Calculate the salvage value of the equipment when it is 4 years old.

   b By how much will the equipment depreciate in the 4 years?

Exercise
4.03

4 Mackenzie borrowed $120 000 to buy a home unit at 6% p.a. monthly reducible interest and her repayments are $1000 per month.

   a How much interest will Mackenzie be charged during the first month?

   b How much will Mackenzie owe at the end of the first month immediately before she makes her first repayment?

   c How much will Mackenzie owe immediately after she makes her first repayment?

   d How much of her first repayment will be interest and how much will go to repaying the loan?

   e How much interest will Mackenzie have to pay during the second month?

   f It will take Mackenzie 15 years and 3 months to repay the loan. Calculate the total amount of interest she will pay.

Use suitable technology to help you answer Questions **5–7**.

Exercises
4.04, 4.05

5 Mackenzie (from Question **4** above) received a $20 000 bonus after she had completed exactly 24 repayments and she decided to use the money to help repay the loan.

   a How much did Mackenzie owe immediately before she repaid a $20 000 lump sum?

   b How much quicker was she able to repay the loan by making the lump sum payment?

   c How much interest did the lump sum payment save her?

6 Rachel borrowed $240 000 from a bank to buy an apartment. The bank offered her an interest rate of 7.2% p.a. monthly reducible for 20 years and told her the monthly repayments would be $1879.

Exercises 4.04, 4.05

  a How much will Rachel repay the bank, including interest?

  b Rachel decided that she could afford to repay $2250 each month. How long will it take her to repay the loan by making the higher monthly repayments?

  c How much interest will Rachel save by making the higher monthly repayments?

7 Beau owes $2050 on his credit card and can only afford to repay the minimum of $55 per month. The credit card charges 22% p.a. Use an online calculator to find:

Exercise 4.07

  a how long it will take Beau to repay his credit card debt if he pays $55 per month

  b how much interest Beau will pay.

5.

FITTING THE DATA

Chapter problem

A consumer association investigated the quality and price of 8 brands of gym shoes, labelled A to H. The investigation team gave each pair of shoes a quality rating out of 10.

	A	B	C	D	E	F	G	H
Price	$320	$280	$260	$240	$180	$180	$80	$40
Quality	8	10	9	6	6	4	1	3

Is there a relationship between the price of shoes and the quality rating assigned by the team? If so, what is the relationship?

CHAPTER OUTLINE

WHAT WILL WE DO IN THIS CHAPTER?

- Draw scatterplots for bivariate data, for example, a person's height and weight
- Describe the association between 2 numerical variables in terms of direction (positive/negative), form (linear/non-linear) and strength (strong/moderate/weak)
- Draw a line of best fit by eye and using technology
- Use the line of best fit to make predictions, both by interpolation and extrapolation

HOW ARE WE EVER GOING TO USE THIS?

- When analysing data to determine whether 2 variables are related, such as height and weight
- When making predictions based on collected data
- When using physical data to design personal training plans such as those used by fitness industry professionals
- The police and security industry analyse data about criminals to predict and prevent crime

A page of scatterplots

Height vs shoe size

Body measurements

5.01 Scatterplots

Bivariate data is data with two variables. For example, you might collect data on the height and weight of a group of students. Bivariate data can be graphed on a **scatterplot**. The first variable is called the **independent variable** and is graphed on the **horizontal axis**. The second variable is called the **dependent variable** and is graphed on the **vertical axis**.

We can look at the scatterplot to see if it has any of the following features:

- there is a pattern or no pattern to the points
- as one variable increases, the other variable increases (or decreases)
- there are groups of points
- most of the points are together but a few are out on their own.

EXAMPLE 1

10 students had their heights and waist heights (above the ground) measured.

Total height, H cm	175	177	178	184	162	172	173	191	163	175
Waist height, W cm	101	106	107	115	100	87	106	120	90	108

a   Graph this bivariate data in a scatterplot.

b   Which is the independent variable and which is the dependent variable?

c   Comment on the features of the scatterplot.

Solution

a   The total height will go on the horizontal axis.

The waist height will go on the vertical axis.

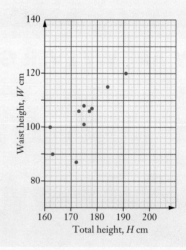

| | b | Independent variable is on the horizontal axis. | The independent variable is height. The dependent variable is waist height. |
| | c | Describe any patterns. | As total height increases, waist height increases. |

Exercise 5.01 Scatterplots

Graph paper is required for this exercise. Keep your scatterplots to use in Exercise 5.02.

1 This table shows the heights of a sample of girls when they were $2\frac{1}{2}$ years old and when they were 18 years old.

Height at $2\frac{1}{2}$ years (cm)	87	83	88	84	82	81	89	86	85	80
Height at 18 years (cm)	171	168	171	167	161	160	174	170	172	158

a Graph this bivariate data in a scatterplot.

b Which is the independent variable and which is the dependent variable?

c Comment on the features of the scatterplot.

2 Matthew asked a group of 11 students how many hours per week they spent playing sport and playing video games.

Sport	6	2	10	4	7	6	10	4	7	5	3
Video games	10	1	0	5	2	12	0	1	3	2	4

  a Present this bivariate data in a scatterplot.
  b Which is the dependent variable and which is the independent variable?
  c Comment on the features of the scatterplot.

3 Simone measured the heights and arm spans of a group of 10 senior students.

Height, H (cm)	170	195	181	181	166	200	163	162	183	167
Arm span, A (cm)	171	186	187	178	165	160	147	143	115	169

  a Construct a scatterplot to show this bivariate data.
  b Which is the independent variable and which is the dependent variable?
  c What does the scatterplot show?

4 The table shows the normal resting pulse of a sample of students and the time it takes each of them to swim 50 m.

Resting pulse (beats/min)	42	70	64	62	55	60	50	72	80
Swimming time (s)		30	48	50	43	40	45	36	49	59

  a Construct a scatterplot for this set of data.
  b Comment on the features of the scatterplot.

5 This table shows the sales of CD albums and digital albums over 9 years. The data is in thousands.

CD albums	46 174	49 818	44 045	38 659	39 529	33 114	30 223	27 356	14 226
Digital albums	91	418	788	1322	2279	3301	4818	6838	7377

  a Present this bivariate data in a scatterplot.
  b Describe what the scatterplot shows.

5.02 What is the relationship?

When bivariate data is graphed on a scatterplot we can use it to see if there is a relationship between the two variables. The **association** between two variables can be described in terms of **direction**, **shape** and **strength**.

The **direction** of the association can be **positive** or **negative**.

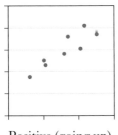

Positive (going up)

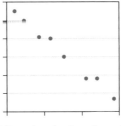

Negative (going down)

The **shape** of the association can be **linear** (in a line) or **non-linear** (in a curve).

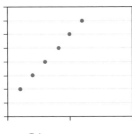

Linear pattern

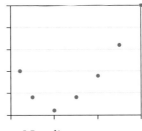

Non-linear pattern

The **strength** of the association can be **strong**, **moderate** or **weak**.

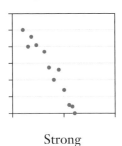

Strong

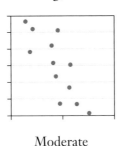

Moderate

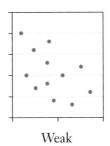

Weak

Sometimes, the points have **no association**.

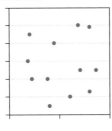

No association

EXAMPLE 2

Describe the association between the variables shown in this bivariate scatterplot.

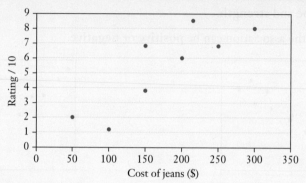

Cost of jeans ($)

Solution

The dots go up from left to right.	Positive
The dots are close to a straight line.	Linear
The points are spread out.	Moderate
	This association is positive, linear and moderate.

Exercise 5.02 What is the relationship?

1 Describe the association (direction, shape and strength) between the two variables shown in each scatterplot.

a

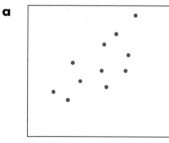

b

c

d

ISBN 9780170413534

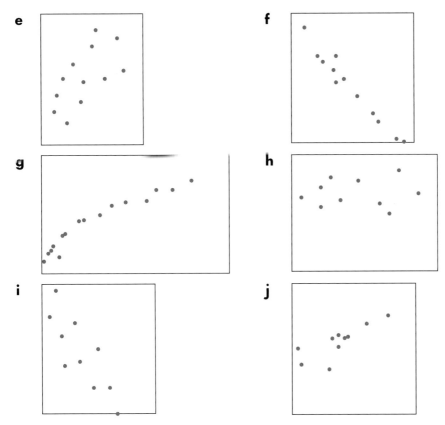

2 Look at the scatterplots you drew in Exercise 5.01. Describe the association of the variables in:

    a   Question **1** on heights   **b**   Question **4** on pulse and time

    c   Question **5** on album sales.

3 The table shows the ages of 10 boys and the time it took them to swim 50 metres.

Age (years)	6	15	7	8	7	9	14	12	10	11
Swim time (s)	76	32	56	49	54	45	36	41	38	36

    a   Construct a scatterplot to display the information.

    b   Does there appear to be a relationship between the boy's ages and the time it takes them to swim 50 m? If so, describe the association.

4 a For each student in your class, record the number of letters in their family name and the number of minutes they spend travelling to school (rounded to the nearest 10 minutes).

b Draw a scatterplot of the data with number of letters on the horizontal axis and travelling time on the vertical axis.

c Would you expect these 2 variables to have any association? Why or why not?

d Is there an association between these 2 variables? If so, describe it.

5 a Collect the following data for each student in your class.
- circumference of the student's head
- student's height

b Draw a scatterplot of the data with circumference of the student's head on the horizontal axis and student's height on the vertical axis.

c Is there a relationship between these two variables? If so, describe it.

6 Sketch scatterplots to illustrate sets of data that have the following features.

a Weak, positive linear association

b Strong, negative linear association

c Strong, positive, non-linear association

d No association

INVESTIGATION

BODY CIRCUMFERENCES

For each student in your class, record the measurements correct to the nearest cm.

- wrist circumference
- neck circumference
- waist circumference

Construct scatterplots for:

- wrist circumference and neck circumference
- wrist circumference and waist circumference

Describe any relationships you can see in each scatterplot.

ISBN 9780170413534

5.03 Looking for a pattern

Let's apply what we have learned to analyse data about a sample of Year 12 students.

Exercise 5.03 Looking for a pattern

This table gives information collected from 15 Year 12 students. You can ask your teacher to download this table as a spreadsheet 'Year 12 data' from NelsonNet.

Year 12 data

Student	Height (cm)	Arm span (cm)	Right foot length (cm)	Time to travel to school (min)	Hours of home-work per week	Resting pulse (beats per minute)	Hours watching TV per week	Number of siblings
Amy	176	175	26	7	6	64	2	4
Joe	178	175	23	40	6	66	10	2
Annika	151	150	30	10	23	95	4	1
Janine	168	175	25	15	7	76	14	2
Stephen	186	181	37	30	23	60	9	3
Thanh	187	187	28	5	2	74	5	2
Gillian	149	149	22	50	17	70	4	1
Vamsee	174	172	28	4	1	62	3	1
Lalaja	172	178	26	20	16	76	20	4
Darryl	177	177	23	15	10	64	10	3
Lyn	169	160	22	7	15	60	0	3
Jeremy	159	155	24	25	9	83	4	2
Ben	169	184	26	34	0	68	17	1
Abdul	163	159	24	20	3	77	9	2
Miriam	163	165	25	20	4	75	1	3

1 Use the data for arm span and length of the right foot.

> You can also use a spreadsheet to draw a scatterplot.

   a Draw a scatterplot for this data.

   b From your graph, name the independent and dependent variables.

   c Describe any features of the scatterplot.

   d Is there an association between these two variables? If so, describe it.

2 Use the data for number of siblings and time to travel to school.

   a Would you expect there to be an association between these two variables? Why or why not?

   b Construct a scatterplot for this data.

   c Describe any features of the scatterplot.

3 Choose 2 variables from the table that you would expect to have an association.

   a Draw a scatterplot for this data.

   b From your graph, name the independent and dependent variables.

   c What features does the scatterplot have?

   d Does the scatterplot show a relationship between these 2 variables? If so, describe it.

4 Choose 2 variables from the table that you would *not* expect to have an association.

   a Draw a scatterplot for this data.

   b Describe any features of the scatterplot.

   c Is there an association between these 2 variables? If so, describe it.

INVESTIGATION

DOES ONE VARIABLE CAUSE THE OTHER?

Just because 2 variables have an association, it doesn't necessarily mean that one causes the other. At one time during the early 2000s there was a strong positive association between the price of petrol and the Australian cricket team's one day run rate! The price of petrol and cricket run rates are unrelated; one couldn't possibly cause the other!

Where we find an association, we need to examine the variables and decide whether or not there is a **causal relationship**. If there is no causal relationship, other factors have to be considered that might cause the variables to have an association.

For example:

- height and weight of a person have a strong positive association. However, a change in height does *not* cause a change in weight. Both height and weight are affected by age and body shape.

- price of petrol and amount of petrol sold have a strong negative association. In this case, one causes the other – as the price increases, sales will decrease because consumers wait for the price to come down again.

Part A

Each of these pairs of variables has a strong association. For each pair:

a decide whether a change in one variable *causes* a change in the other variable

b if one causes the other, explain how

c if there is no causal relationship, suggest other factors that might cause the variables to have an association.

   1 Driving speed of a car and amount of petrol used

   2 Length of the right foot and length of the left foot of the same person

   3 Price of a particular brand of car and number of cars sold

4 Sales of hot chips and soft drinks at a football game

5 Height and arm span of students

6 Calories consumed by a person and weight gained

7 Number of rainy days in a month and sales of umbrellas in that month

Part B

a Write 2 variables you would expect to have a positive linear relationship where a change in one variable would cause a change in the other variable.

b Write 2 variables you would expect to have a positive linear relationship where a change in one variable would *not* cause a change in the other variable.

c Write 2 variables you would expect to have a negative linear relationship where a change in one variable would cause a change in the other variable.

5.04 Drawing a line of best fit

If the data on a scatterplot shows a strong linear association, we can approximate the linear relationship by drawing a **line of best fit** through the points.

A line of best fit:

- represents most or all of the points as closely as possible

- goes through as many points as possible

- has roughly the same number of points above and below it

- is drawn so that the distances of points from the line are as small as possible.

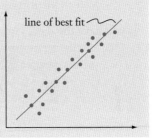

line of best fit

EXAMPLE 3

This table from page 131 shows the heights of a sample of girls at ages $2\frac{1}{2}$ and 18.

Height at $2\frac{1}{2}$ years (cm)	87	83	88	84	82	81	89	86	85	80
Height at 18 years (cm)	171	168	171	167	161	160	174	170	172	158

Graph this bivariate data on a scatterplot and draw a line of best fit for it.

Solution

Graph the data first.
Then, draw a line through the middle of the points with roughly the same number of points above and below the line.

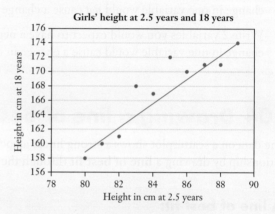

If you have the data, you can use a spreadsheet to draw the line of best fit.

1   Enter the data into the spreadsheet, then graph it on a scatterplot.

2   Select the graph and select **Trendline** from the **Layout** menu.

3   Select **Linear Trendline** and the line will appear on the graph.

Exercise 5.04 Drawing a line of best fit

Keep your answers to this set of exercises. You will be using them again in Exercise 5.06.

1 Copy or print each scatterplot and draw a line of best fit. You can print a copy of the graphs from the worksheet 'Scatterplots' from NelsonNet.

a

b

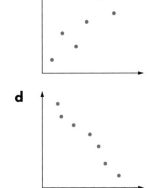

c

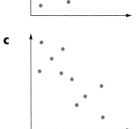

d

2 This table shows birth rate and female life expectancy for a number of countries. Birth rate is births per year per 1000 population. Life expectancy is measured in years.

Country	Birth rate	Female life expectancy
Australia	13.3	85.5
Brazil	14.7	79.8
Canada	11.2	84.6
Fiji	21.4	76
Germany	8.1	83.5
Iraq	31.0	72
Kenya	36.1	61
Laos	28.0	69.5
Nepal	24.3	70
Niger	50.0	57
Rwanda	42.1	61

a Draw a scatterplot of birth rate against life expectancy.
b Draw a line of best fit for this data.

3 This table shows the height above sea level and the average annual rainfall for some locations in and around Australia.

City	Height above sea level (m)	Mean annual rainfall (mm)
Alice Springs	581	282
Ballarat	432	694
Hobart	24	616
Kalgoorlie	387	266
Mount Isa	365	463
Norfolk Island	73	1017
Perth	15	736
Winton	188	382

a Draw a scatterplot of height above sea level against mean annual rainfall.

b Draw a line of best fit for this data.

4 This table shows information about the amount of energy, carbohydrate and fat contained in 100 g of some take-away foods.

Food	Energy (kilojoules)	Carbohydrate (grams)	Fat (grams)
Hamburger	1030	26.6	9.1
Cheeseburger	1070	23.9	11.2
Chicken burger	921	20.2	10.1
Fish burger	988	24.5	10.2
Grilled chicken wrap	771	13.4	10.2
Chicken salad	325	5.5	3.4
Egg and bacon wrap	767	15.3	9.1
Hash browns	1150	26.4	17.2
Chips	1480	39.7	19.0

a Draw a scatterplot of energy against carbohydrate.

b Draw a line of best fit for this data.

c Draw a scatterplot of energy against fat.

d Draw a line of best fit for this data.

NCM 12. Mathematics Standard 1

ISBN 9780170413534

5.05 Using a line of best fit

Lines of best fit show the association between variables and we can use them to make predictions within the range of the data.

Lines of fit

EXAMPLE 4

Scott is a competitive swimmer. He noticed that his times for the 100 m freestyle event are slower when he smokes cigarettes in the 3 days before the event. On a scatterplot, he displayed his times for his last 8 events and the numbers of cigarettes he smoked in the 3 days before each event. He drew a line of best fit through the data.

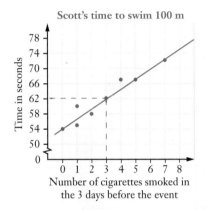

a   What is the independent variable?

b   What is the dependent variable?

c   What does the line of best fit show about the relationship between these 2 variables?

d   Use the line of best fit to predict Scott's time for the 100 m freestyle when he has smoked 3 cigarettes in the 3 days before the event.

Solution

a   Independent variable is on the horizontal axis.

The independent variable is the number of cigarettes smoked.

b   Dependent variable is on the vertical axis.

The dependent variable is the swimming time.

c   Describe the relationship.

As the number of cigarettes smoked increases, the time for the 100 m increases.

d   Find 3 on the horizontal axis, go up to the line and then across to the time.

Scott's time will be approximately 62 seconds.

1 Anita is trying to strengthen her quadriceps muscles. This graph shows the number of leg extensions she was able to complete in 30 seconds each week after she started training.

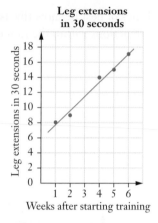

Leg extensions in 30 seconds

a What is the independent variable?

b What is the dependent variable?

c What does the line of best fit show about the relationship between these 2 variables?

d Anita forgot to record the number of leg extensions she could do in week 3. Use the line of best fit to predict the number.

ISBN 9780170413534

2 Humanitarian agencies send people to assist when a disaster occurs. This scatterplot shows the number of medical staff one small agency sent to assist after earthquakes and floods occurred in different parts of the world and the estimated number of victims affected by each disaster.

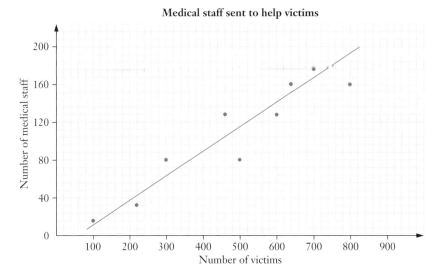

Medical staff sent to help victims

a What is the independent variable?

b What is the dependent variable?

c Use the line of best fit to estimate the number of medical staff the agency will send to a disaster with an estimated 200 victims.

d The agency sent 180 medical staff to assist in a disaster. Approximately how many victims were affected by the disaster?

e Describe the relationship between the number of medical staff and the number of victims.

f Kyle noted that the greater the number of medical staff, the greater the number of victims. He argued that if the agency decreased the number of medical staff, there would be fewer victims. What is wrong with Kyle's argument?

3 Billy announced, 'People with big feet are better at maths than people with small feet'. Billy gave a short Maths test to all the people in the school library and measured the length of each person's right foot. He displayed the results on a scatterplot.

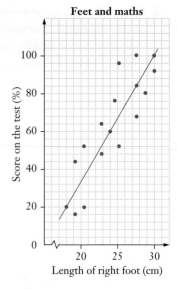

Feet and maths

a According to Billy's line of best fit, what is the right foot length of a person who scored 58 on Billy's maths test?

b Describe the relationship between the length of the right foot and the score on the test as shown by this line of best fit.

c Big feet don't cause high maths scores. How can Billy's results be explained?

4 This graph shows the scatterplot for the heights and weights of 20 students with a line of best fit drawn on it.

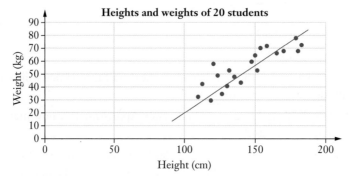

Heights and weights of 20 students

a Describe the relationship between the height and weight of students as shown by this line of best fit.

b Use the line of best fit to estimate the weight of a student who is 150 cm tall.

c Does an increase in height *cause* an increase in weight? Why or why not?

NCM 12. Mathematics Standard 1 ISBN 9780170413534

5 Daniel graphed the number of sunny days in January against the number of dry cloudy days in January for 12 Australian locations. He then drew a line of best fit on his scatterplot.

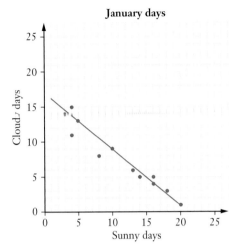

January days

a Use the line of best fit to estimate how many dry cloudy days there would be if there were 7 sunny days.

b Describe the relationship between the number of sunny days and the number of cloudy days in January.

c Suggest a reason for this relationship.

5.06 Interpolation and extrapolation

A line of best fit helps us to make **predictions** as we did in some of the questions in Exercise 5.05. If we make a prediction on information that lies *within* the data, this is called **interpolation**. If we make a prediction on information that lies *outside* the data, this is called **extrapolation**.

However, we need to be careful when we are making predictions from sets of data.

- Predictions within the data can be unreliable if the association is weak. Predictions are more reliable if the association is strong and linear.

- Predictions that go outside the data can be unreliable because we can't be sure that the linear relationship continues beyond the data.

EXAMPLE 5

In a science experiment, Clare and Dilani attached different weights to a spring. They recorded the length of the spring for each weight.

Mass (g)	100	200	300	400	500
Length of spring (mm)	26	33	47	59	65

They graphed the results on the scatterplot below and drew a line of best fit.

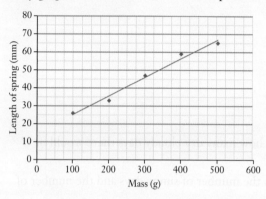

a   Use the line to find:

   i   the length of the spring for a mass of 250 g

   ii   the mass when the length of the spring is 30 mm.

b   How reliable are these predictions?

c   The equation of the line is $L = 0.104m + 14.8$. Use it to predict the length of the spring when the attached mass is 800 g.

d   How reliable is this prediction?

Solution

a   i   Find 250 g on the horizontal axis, go up to the graph and across to length on the vertical axis.

The length of the spring is approximately 40 mm.

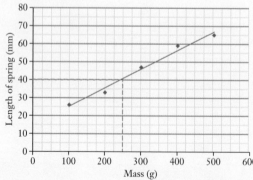

ii Find 30 mm on the vertical axis, go across to the line and down to mass on the horizontal axis.

The mass is approximately 150 g.

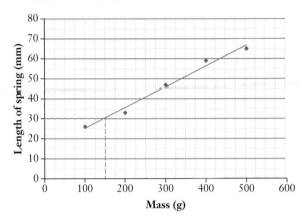

b These predictions are within the given data.

These predictions are reasonably reliable as the association is strong and linear.

c Substitute $m = 800$ into the formula $L = 0.104m + 14.8$.

$L = 0.104 \times 800 + 14.8$
   $= 98$
The length of the spring with 800 g attached is approximately 98 mm.

d A mass of 800 g is outside the data shown on the graph, so it is an example of extrapolation.

The prediction is not necessarily reliable as we don't know if the relationship continues to be linear. It is also possible that at some point the spring might break.

Exercise 5.06 Interpolation and extrapolation

This exercise requires some of your answers from Exercise 5.04.

1 Jason recorded the body mass of a sample of women at the gym and the amount of energy they use per day.

Body mass (kg)	60	70	75	82	89	95
Energy used (kJ)	1040	1140	1190	1260	1300	1340

He graphed the data on a scatterplot and drew a line of best fit.

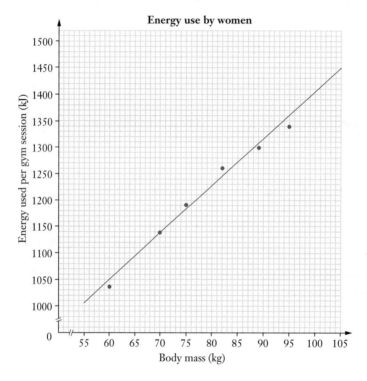

a Use the line of best fit to find:

i the amount of energy used by a woman of mass 65 kg

ii the body mass of a woman who uses 1450 kJ of energy per day.

b How reliable are these predictions? Justify your answer.

c The equation of the line is $E = 16.9B - 143$, where B is body mass and E is energy used. Use it to predict the energy used by a woman of mass 120 kg.

d Would you expect this prediction to continue to be a linear relationship beyond the data points? Why or why not?

NCM 12. Mathematics Standard 1   ISBN 9780170413534

2 In Exercise 5.04, Question **2** you drew a scatterplot for data on the birth rate (B) and female life expectancy (E) in a number of countries. A line of best fit for this data has equation $E = -0.725B + 91.2$.

   a Predict the female life expectancy in a country where the birth rate is 40.

   b Determine the birth rate in a country where female life expectancy is 65 years.

   c How reliable are these predictions? Justify your answer.

   d What would be the female life expectancy in a country where the birth rate is 60?

   e How reliable do you think estimates outside this data would be? Justify your answer.

3 In Exercise 5.04, Question **3** you drew a scatterplot for data on the height above sea level (H) and mean annual rainfall (R) for a number of locations. A line of best fit for this data has equation $R = -0.769H + 755.5$.

   a Predict the mean annual rainfall for a city 300 m above sea level.

   b Determine the height above sea level for a city that receives 400 mm of rain per year.

   c How reliable are these predictions? Justify your answer.

   d Calculate the mean annual rainfall for a city 800 m above sea level.

   e Predict the mean annual rainfall for a city 1000 m above sea level. Is this possible? What is wrong with this prediction?

4 Ms Cranston has a set of results for her senior class on two Maths tests.

Student	Bill	Ruth	Mary	Ella	Greg	Jim	Meg	Tara	Bob	Clem	Bree	Amy
Algebra test	60	38	65	??	75	48	67	23	82	16	92	80
Data test	35	21	47	31	56	40	54	11	59	20	62	??

Ella and Amy missed a test and Ms Cranston wants to give them an estimate.

   a Draw a scatterplot for this data (except Ella and Amy) and add a line of best fit.

   b Predict the results that Ella and Amy might have received in the test they missed.

   c How reliable do you think these predictions are? Justify your answer.

5 Ryan investigated the heights (x cm) of a group of his friends and their hand spans (y cm, the maximum distance between their thumb and little finger).

Height (x cm)	170	178	160	183	168	145	155
Hand span (y cm)	20	21	19	22	20	17	19

   a Construct a scatterplot for the data and draw a line of best fit on the plot.

   b A line of best fit has equation $y = 0.12x - 0.02$. Use the equation to predict the hand span of Izak who is 165 cm tall. Answer to the nearest centimetre.

   c Robert Wadlow was the tallest man in the world. His height was 272 cm. Use the equation of the line of best fit to predict his hand span correct to the nearest cm, and explain why this measurement is unlikely to be correct.

KEYWORD ACTIVITY

COMPLETE THE BLANKS

Use the listed words below to copy and complete a summary of this chapter.

association	bivariate	dependent variable
extrapolation	independent variable	interpolation
linear	line of best fit	moderate
negative	non-linear	positive
predictions	scatterplot	strong	weak

In this chapter, we studied **1**_____ data, which is data with two variables. We have learned how to graph this data on a **2**_____. On the graph, the variable on the horizontal axis is called the **3**_____ _____ and the variable on the vertical axis is called the **4**_____ _____.

We can use the graph to decide whether or not there is an **5**_____ between the variables. We considered this relationship in terms of its:

Direction: whether it is **6**_____ or _____.

Shape: is it **7**_____ or _____?

Strength: is the relationship **8**_____, _____ or _____?

We then added a **9**_____ __ ____ ___ to the scatterplots.

We can use the lines to help us make **10**_____ from the data. When we make estimates from within the data range this is called **11**_____ and is usually reliable if the association is strong. When we make estimates outside the range of the data, this is called **12**_____ and can often be unreliable or impossible.

SOLUTION TO THE
CHAPTER PROBLEM

Problem

A consumer association investigated the quality and price of 8 brands of gym shoes, labelled A to H. The investigation team gave each pair of shoes a quality rating out of 10.

	A	B	C	D	E	F	G	H
Price	$320	$280	$260	$240	$180	$180	$80	$40
Quality	8	10	9	6	6	4	1	3

Is there a relationship between the price of shoes and the quality rating assigned by the team? If so, what is the relationship?

Solution

Draw a scatterplot to see whether there is a relationship.

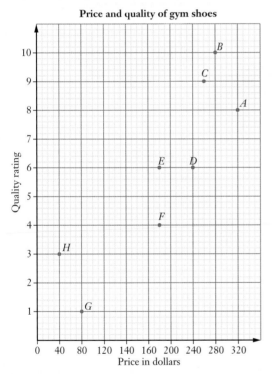

The scatterplot shows there is a relationship between the 2 variables. Overall, the more expensive the gym shoe, the better the quality.

The relationship is strong, positive and linear. You would expect that the more expensive shoes use better materials and are better constructed than the cheap shoes.

If you wish, you could draw a line of best fit and use it to make predictions about other pairs of shoes.

- Have you learned anything new in this topic? If so, what?
- Give some examples of where the skills from this chapter might be used.
- Is there anything you didn't understand? If so, ask your teacher for help.

Copy and complete this mind map of the topic, adding detail to its branches and using pictures, symbols and colour where needed. Ask your teacher to check your work.

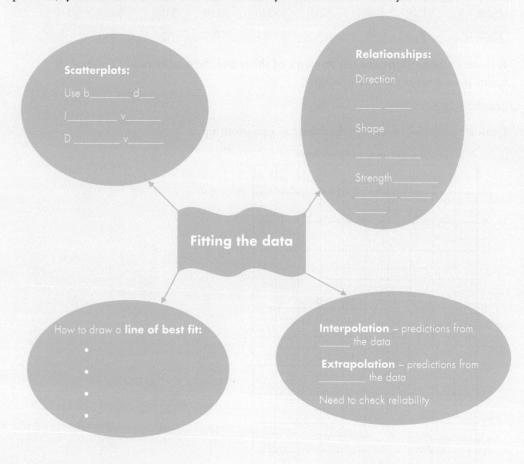

Scatterplots:

Use b_____ d____

l_____ v_____

D _____ v_____

Relationships:

Direction

_____ _____

Shape

_____ _____

Strength_____

_____ _____

_____

Fitting the data

How to draw a **line of best fit:**

-
-
-
-

Interpolation – predictions from _____ the data

Extrapolation – predictions from _____ the data

Need to check reliability

5. TEST YOURSELF

1 Eliza works in a coffee shop. She thinks there is a relationship between the outside temperature and their daily hot chocolate sales.

Temperature (°C)	8	36	12	28	14	22	16	10	24	18	7	31
Number of hot chocolate drinks sold	42	15	37	24	9	20	35	20	10	30	30	4

  a Graph Eliza's bivariate data in a scatterplot.

  b Which is the independent variable and which is the dependent variable?

  c Comment on the features of the scatterplot.

2 Describe the association (direction, shape and strength) between the 2 variables shown in the scatterplot you drew in Question **1**.

Exercise 5.02

3 Sketch scatterplots to illustrate sets of data that have the following features.

  a Weak, negative linear association

  b Strong, positive linear association

  c No association

4 This data comes from the table on page 137.

  a Do you expect to find a strong association between these 2 variables? Why or why not?

  b Draw a scatterplot for this data.

  c From your graph, name the independent and dependent variables.

  d Is there an association between these 2 variables? If so, describe it.

Student	Hours of homework per week	Hours watching TV per week
Amy	6	2
Joe	6	10
Annika	23	4
Janine	7	14
Stephen	23	9
Thanh	2	5
Gillian	17	4
Vamsee	1	3
Lalaja	16	20
Darryl	10	10
Lyn	15	0
Jeremy	9	4
Ben	0	17
Abdul	3	9
Miriam	4	1

Exercise
5.04

Exercise
5.05

Exercise
5.06

5 Refer to the graph you drew in Question **1a**. Draw a line of best fit for this data.

6 Use the line of best fit that you drew in Question **5** to answer the following questions.

   a How many hot chocolate drinks could Eliza expect to sell on a day with a temperature of 20°C?

   b Eliza sells 25 hot chocolate drinks. Predict the temperature on this day.

7 Mr Armstrong, the Science teacher, sets up an experiment to measure the pressure of a gas at different temperatures. This table shows the results.

Temperature (T°C)	20	30	40	50	60	70	80	90
Pressure (P g/cm^3)	30	29.8	32.1	31.9	34.1	33.8	34.8	36.6

   a Draw a scatterplot for this data.

   b Draw a line of best fit on your scatterplot.

   c Use the line of best fit to estimate the pressure at 56°C.

   d How reliable is this estimate? Justify your answer.

   e A line of best fit for this data has the equation $P = 0.094T + 27.748$. Use the equation to predict the pressure at −10°C.

   f How reliable is this estimate?

Practice set 1

Section A: Multiple-choice questions

For each question, select the correct answer **A**, **B**, **C** or **D**.

1 Which of the following ratios is not equivalent to 32 : 48?

   A   16 : 24   **B**   4 : 6   **C**   2 : 3   **D**   6 : 8

Exercise 2.01

2 Which equation has a gradient of 3?

   A $y = x + 3$ **B** $y = 3 - x$ **C** $y = 3x$ **D** $y = 3$

Exercise 3.01

3 Sunil receives a 7% pay rise on a salary of $67 000. What is his new salary?

   A $67 007   **B** $67 469   **C** $71 690   **D** $113 900

Exercise 1.01

4 Andrea owes $370 on her credit card. She makes a payment of $250. While shopping, she charges a further three items to her card costing $35, $77 and $24. How much does she now owe on her credit card?

   A $256   **B** $386   **C** $484   **D** $756

Exercise 4.06

5 Which scatterplot shows an association that is moderate, linear and positive?

Exercise 5.02

A

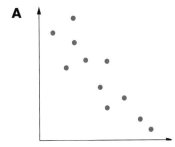

C

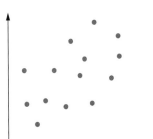

B

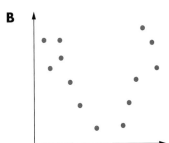

D

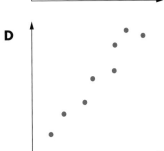

Exercise
1.02

6 Calculate the simple interest earned on $800 at 4% p.a. for 3 years.

   A $12   **B** $24   **C** $32   **D** $96

Exercise
2.03

7 In a mining town, the ratio of women to men is 2 : 5. There are 240 women. Altogether, how many men and women are there in the town?

   A   600   **B**   840   **C**   1200   **D**   1680

Exercise
3.03

8 What is the point of intersection of the two lines shown in this graph?

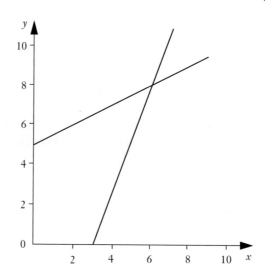

   A   (0, 5)

   B   (3, 0)

   C   (8, 6)

   D   (6, 8)

Exercise
1.04

9 Jane invests $4500 at 3.5% p.a. compounding annually for 8 years. Calculate the amount of interest she earns.

   A $1260   **B** $1425.64   **C** $5925.64   **D** $9000

Exercise
2.04

10 A street map has a scale of 1 : 20 000. What distance on the map represents 9.6 km?

   A   480 cm   **B**   48 cm   **C**   4.8 cm   **D**   0.48 cm

11 Goran buys a new car for $30 000 that depreciates by 18% p.a. By how much has the car depreciated over the first 3 years?

Exercise 4.02

    A $5400   **B** $13 459   **C** $16 200   **D** $16 541

12 This graph shows the income and expenditure for Catriona's Cupcakes.

Exercise 3.04

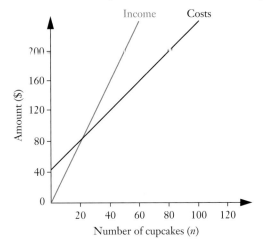

What happens when Catriona sells 40 cupcakes?

    A   She makes a profit

    B   She makes a loss

    C   She breaks even

Section B: Short-answer questions

Exercise 4.01

1 Thomson Tiles bought a new office computer system for $11 400. The system depreciates by $2050 each year.

 a By how much will the computer system depreciate in 3 years?

 b Calculate the system's salvage value at the end of 3 years.

 c After how long will the system be worth nothing?

Exercise 2.02

2 For their new bathroom, Ann and Chris need 4 green tiles for every 3 white tiles. They buy 100 green tiles. How many white tiles do they need to buy?

Exercise 1.04

3 Rashid invested $3500 at 5.9% p.a. compounding annually for 3 years.

 a Calculate the value of Rashid's investment at the end of the 3 years.

 b How much interest did Rashid earn?

Exercise 2.03

4 At Nelson Valley High School, the ratio of students who speak a second language to students who speak English only is 5 : 8. There are 923 students in the school. How many students speak English only?

Exercise 3.02

5 Year 12 is running a sausage sizzle at the athletics carnival. It will cost them $32 for the gas bottle for the barbecue and each sausage sandwich will cost $0.90 to make.

 a Copy and complete this table.

Number of sausage sandwiches, n	0	10	20	30	40	50	100	150	200
Cost, $C	32	41							

 b Write a linear function to calculate the cost of making n sausage sandwiches.

 c Use the function to calculate the cost of making 170 sausage sandwiches.

 d Construct a graph showing the cost of making n sausage sandwiches.

 e Find the vertical intercept and explain what this value represents.

 f Find the gradient and explain what this value represents.

6 Year 12 decides to sell the sausage sandwiches for $2 each.

Exercise 3.04

a Copy and complete this table.

Number of sausage sandwiches, n	0	10	20	30	40	50	100	150	200
Income, $I	0	20							

b Write a linear function to calculate the income from selling n sausage sandwiches and graph it on the same axes you used in Question **5**.

c What are the coordinates of the point where the two lines intersect?

d What does this point represent?

e Will Year 12 make a profit if they sell 70 sausage sandwiches? Explain your answer.

f How much profit will Year 12 make if they sell all 200 sausage sandwiches?

7 A new sound system for Zengage DJs costs $6250 and it is depreciating at 20% per year. Calculate the sound system's salvage value when it is 4 years old.

Exercise 4.02

8 Juanita collected some data from students at her school about how many hours per week they spent playing sport and watching TV.

Exercises 5.01, 5.02

Sport	5	6	17	9	3	13	17	5	4	0	10
TV	8	6	1	2	10	0	2	9	9	10	2

a Present this bivariate data in a scatterplot.

b Comment on the features of the scatterplot.

c Describe the association between the variables.

9 a Draw a line of best fit for the data in your scatterplot from Question **8**.

Exercises 5.04, 5.05, 5.06

b What does the line of best fit show about the relationship between these two variables?

c Use the line of best fit to predict the hours of TV watched by a person who plays 8 hours of sport.

d Use the line of best fit to predict the hours of sport played by a person who watches 3 hours of TV.

e How reliable are these predictions? Justify your answer.

Exercise
2.04

10 Use the map of Nambucca Heads in Test yourself 2, Question **11** on page 59.

   a Find the distance between the Cemetery and Coronation Park.

   b Find the length of the Causeway.

   c Aaron walks 3 km each morning before work.

      i What distance will this be on the map?

     ii If he walks from the Cemetery south around the border of the State Forest, will he walk 3 km?

Exercise
3.03

11 a Graph the lines $x + y = 5$ and $2x - y = 7$ on the same set of axes. You may use tables of values or the technology of your choice.

   b Find the solution of the simultaneous equations $x + y = 5$ and $2x - y = 7$.

Exercise
1.04

12 Oliver invests $8400 for 18 months at 2.4% p.a. compounding monthly.

   a What is the value of the investment at the end of 18 months?

   b How much interest did Oliver earn?

Exercise
4.03

13 Rose and Ian take out a personal loan to pay for an overseas holiday. They borrow $18 000 at 8.4% p.a. with monthly repayments of $475. Copy and complete the table below showing the first 4 months of their loan.

Rose and Ian's reducible finance loan				
Amount borrowed: $18 000				
Interest rate: 8.4% p.a. monthly reducing				
Monthly repayments: $475				
Month	Principal (P)	Interest (I)	Principal + interest ($P + I$)	Amount owing ($P + I - R$)
1st	$18 000	$126	$18 126	$17 651
2nd				
3rd				
4th				

14 On a tourist map of Sydney, the scale used is 1 : 9500.

   a   Find the actual distance between the following places given the scaled distance:

      i   Circular Quay station to the Museum of Contemporary Art (2.2 cm)

      ii one end of Martin Place to the other (5 cm).

   b   Find the scaled distance between the following places given the actual distance:

      i   Central Station to the Capitol Theatre (294.5 m)

      ll Pyrmont Bridge to the Art Gallery of NSW (1.4 km).

15 Ethan earns a salary of $72 300. Assuming inflation is 2.1% p.a., calculate Ethan's salary in 5 years. Answer to the nearest dollar.

16 Jason owes $4720 on his credit card, which has a flat interest rate of 19.99% p.a.

   a   How much interest will Jason be charged for 1 month?

   b   Calculate the minimum repayment, which is 2% of the amount owed.

   c   How much will Jason still owe on his credit card after he makes the minimum repayment?

ALGEBRA

6.

GRAPHING CURVES

Chapter problem

The number of people living on New Century Road is increasing by 9% p.a. and currently there are 90 people living there.

a Complete a table to show its population in the first 5 years. In your calculations, round to the nearest whole number.
b Draw a graph showing this population growth.
c The equation for calculating the population, P, after t years is given by $P = 90 \times (1.09)^t$. Calculate the population of New Century Road in 10 years.
d Suggest factors that could complicate the modelling of this real-life situation.

CHAPTER OUTLINE

WHAT WILL WE DO IN THIS CHAPTER?

- Graph functions that give a curve, not a straight line, including the quadratic and exponential functions
- Work with applications of quadratic and exponential functions, such as population growth
- Draw a graph of a practical situation that doesn't follow a formula, such as the noise level of a classroom over a lesson period

HOW ARE WE EVER GOING TO USE THIS?

- When interpreting data that is non-linear
- When designing satellite dishes, bridges, roller coasters and car headlights, as parabolas are the basis of their design
- When modelling and graphing the growth of investments or populations

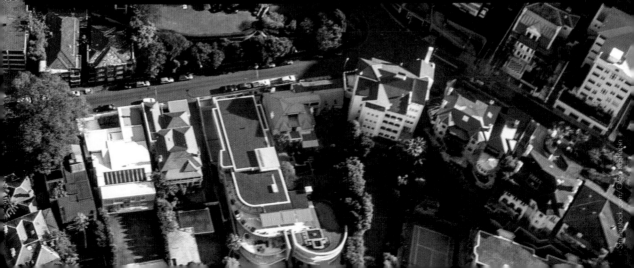

6.01 Graphing tables of values

Some functions are not linear. In this chapter, we will graph some data that gives us curves rather than lines. A **non-linear function** is a function whose graph is *not* a straight line.

Isobel plays basketball. When she attempts to shoot a goal the ball follows a curve. This table gives the height, *h* metres, of the ball *t* seconds after it leaves her hands.

Time, *t* (seconds)	0	0.5	1	1.5	2	2.5
Height, *h* (metres)	2.4	3.3	3.7	4.0	3.6	2.8

Draw the graph of this data and join the points with a smooth curve.

Solution

Draw a set of axes with time on the horizontal axis and height on the vertical axis. Plot the points. Join them with a smooth curve.

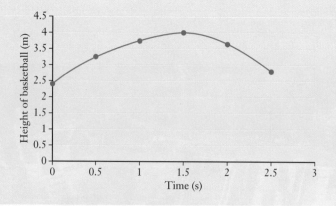

Exercise 6.01 Graphing tables of values

1 A half-pipe is included in the new skate park at Nelsonville Village Park. This table gives the shape of the half-pipe.

Example 1

Distance from left-hand side (metres)	0	1	2	3	4	5	6	7	8
Height of pipe (metres)	2.4	1.35	0.6	0.15	0	0.15	0.6	1.35	2.4

Graph the table of values and join the points with a smooth curve.

2 This table shows the height of the tides in Cairns over 3 consecutive days.

Day	Wednesday				Thursday				Friday			
Time	1 a.m.	7:30 a.m.	1:15 p.m.	7:45 p.m.	1:45 a.m.	8 a.m.	2 p.m.	8:15 p.m.	2:15 a.m.	8:45 a.m.	2:30 p.m.	8:45 p.m.
Height of tide (m)	0.5	2.6	0.6	2.8	0.4	2.7	0.6	2.7	0.4	2.8	0.8	2.6

a Graph the table of values as a smooth curve.

b Use your graph to estimate the height of the tide at midnight Wednesday.

c Joe needs a tide height of 1 m to go fishing. Use your graph to estimate when he can go fishing on Friday afternoon.

3 For a Science project, Ellie measures the temperature of a cup of coffee at different times after she makes it. Her data is given in the table below.

Time (min)	0	4	8	12	16
Temperature (°C)	80	68	58	49	42

a When you draw this graph, should Time or Temperature be on the horizontal axis?

b Draw the graph of this data and join the points with a smooth curve.

c Use your graph to estimate the temperature of the coffee after 2 minutes.

d Ellie likes to drink her coffee at approximately 53°C. After she makes her coffee how long should she wait to drink it?

4 These tables show the hours and minutes of daylight in 2 cities on the 21st day of the month over one year. For example, 6:52 means 6 hours 52 minutes of daylight.

Anchorage, Alaska											
Jan	**Feb**	**Mar**	**Apr**	**May**	**Jun**	**Jul**	**Aug**	**Sep**	**Oct**	**Nov**	**Dec**
6:52	9:39	12:20	15:18	17:59	19:22	18:00	15:16	12:21	9:31	6:47	5:27

Perth, Australia											
Jan	**Feb**	**Mar**	**Apr**	**May**	**Jun**	**Jul**	**Aug**	**Sep**	**Oct**	**Nov**	**Dec**
14:39	13:25	12:07	10:43	9:37	9:09	9:36	10:42	12:05	13:27	14:40	15:11

a On the same set of axes, graph the data for each city and join the points with a smooth curve. Use a different colour for each city.

b What are the similarities between the 2 graphs?

c What are the differences between the 2 graphs?

d In which months do the graphs cross?

e Why is the maximum hours of daylight in Anchorage in June greater than the maximum hours of daylight in Perth in December?

5 Max is organising a bus trip to see a concert. The cost of the bus is $840 and the maximum number of people it carries is 40.

a Copy and complete this table to show the cost per person for different numbers of people going by bus to the concert.

Cost per person = 840 ÷ the number of people

Number of people	1	5	10	20	30	40
Cost per person ($)	840	168				

b Graph this data, joining the points with a smooth curve.

c From your graph, estimate the cost per person if 28 people attend.

d Calculate the cost using the formula. How close was your estimate from the graph?

6.02 Quadratic modelling

Tables of values

Graphing
quadratic
functions

Graphing
curves: graphics
calculator

A **quadratic function** is a function where the highest power of x is x^2, for example, $y = 2x^2$ and $y = -x^2 + 9$.

In Example 1 and in Exercise 6.01, Question 1 we graphed curves shaped like this:

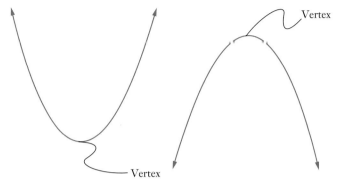

This type of curve is called a **parabola**, the graph of a quadratic function. It is symmetrical with a turning point called a **vertex**.

Parabolas are everywhere! Every satellite dish and car headlight has the shape of a parabola. When we throw a ball, fire a cannon or spurt water out of a hose, quadratic functions come into play.

EXAMPLE 2

Quadratic
functions

A rock fell from the top of a cliff onto a road. The graph shows the height of the rock above the road t seconds after it fell.

a How high was the rock above the road after 2 seconds?

b How far did the rock fall during those first 2 seconds?

c Approximately after how many seconds did the rock hit the road?

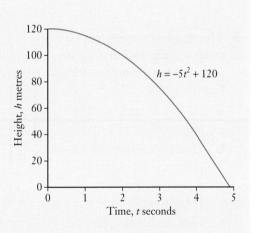

Solution

a Read from the graph.

The rock was 100 m above the road after 2 seconds.

b The rock was 120 m high before it fell and 100 m high after 2 seconds.

During the first 2 seconds it fell:
120 m – 100 m = 20 m

c Read from the graph.

The rock hit the road at about 4.9 s.

Exercise 6.02 Quadratic modelling

1 When Jose jumps out of a plane, he free-falls for 15 seconds before he pulls his parachute cord. The graph shows his height above the ground during the free-fall.

   a At what height did Jose jump out of the plane?

   b How high was Jose above the ground 10 seconds after he jumped?

   c How far did Jose fall in the first 10 seconds?

   d Approximately how high above the ground was Jose when he pulled his parachute chord?

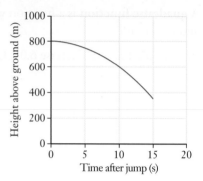

2 Adriana is planning to make a square chicken shed on her farm. The graph shows the relationship between the length and area of the shed.

Use the graph to estimate the length of the chicken shed with an area of 30 m².

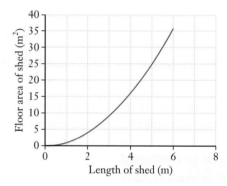

3 This graph represents part of a roller coaster track at a theme park. It is based on a quadratic function.

   a At what height does this section of the ride start?

   b How high above the ground is the ride when it has travelled 50 m horizontally?

   c How far has the ride travelled horizontally when it reaches this height again?

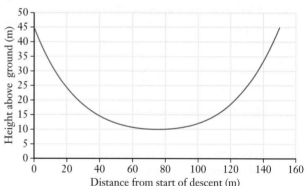

4 Naresh hit a cricket ball out of the oval. The graph shows the path of the ball in metres.

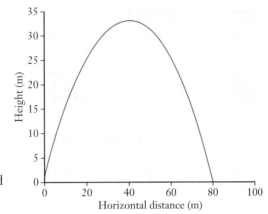

a Estimate the maximum height of the ball.

b Approximately how far from Naresh did the ball land?

c While falling, the ball just cleared the top of the scoreboard, which is 12 m high. How far is the scoreboard (horizontally) from Naresh?

5 A satellite dish is built using the formula $h = 0.05d^2$, where h is the height above the base of the dish and d is the horizontal distance from the centre of the dish.

a Copy and complete this table of values for the equation.

Horizontal distance, d m	−32	−24	−16	−8	0	8	16	24	32
Height, h m	51.2								

b Graph $h = 0.05d^2$.

c How high above the centre are the sides of the dish?

d Estimate the horizontal distances from the centre to where the dish is 20 m high.

6 One half of the advertising logo for Big M is created using the equation $h = -d^2 + 8d$, where h is the height of the curve and d is the horizontal distance from the start.

a Copy and complete this table of values for the equation.

Horizontal distance, d m	0	1	2	3	4	5	6	7	8
Height of curve, h m									

b Graph $h = -d^2 + 8d$.

c What is the height of the centre of one half of the M?

d Estimate the horizontal distances from the start of this half of the M to where the height will be 8 m.

7 A rectangle has length x cm and width $18 - x$ cm.

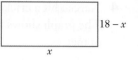

$18 - x$

x

a Use algebra to show that the perimeter of this rectangle is 36 cm.

b When $x = 10$, what are the dimensions of the rectangle?

c What is the area of the rectangle when $x = 10$ cm?

d Explain why the value of x can't be negative or larger than 18.

e Explain why $A = 18x - x^2$ represents the area of a rectangle with a perimeter of 36 cm.

f Copy and complete this table of values.

Length of side, x cm	0	2	4	5	7	9	11	13	15	17	18
Area of rectangle, A cm^2											

g Graph $A = 18x - x^2$.

h What is the maximum area a rectangle with a perimeter of 36 cm can have?

i What special shape is the rectangle with the maximum area?

j Use technology to graph the function $A = 18x - x^2$ and compare the result to the parabola you constructed.

6.03 Exponential growth and decay in finance

When a quantity increases repeatedly by a percentage of itself, we say that the quantity is growing exponentially. In Chapter 1, we learned about compound interest and inflation. These are examples of **exponential growth**.

When we graph a quantity that is growing exponentially, the graph looks like this:

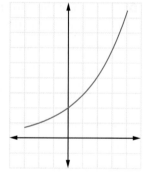

When a quantity decreases repeatedly by a percentage of itself, we say that the quantity is decaying exponentially. In Chapter 4, we learned about the declining-balance method of depreciation. This is an example of **exponential decay**.

When we graph a quantity that is decaying exponentially, the graph looks like this:

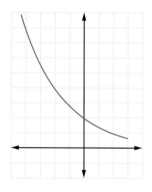

EXAMPLE 3

Antonio invested \$2700 for 5 years at 4.5% p.a. compounding annually.

a Complete this table of values for the value of the investment over 5 years.

Year	0	1	2	3	4	5
Value of investment (\$)	2700					

b Draw a graph showing the value of his investment over 5 years.

Solution

a Use the compound interest formula $FV = PV(1 + r)^n$ to complete the table. Round answers to the nearest dollar.

Year	0	1	2	3	4	5
Value of investment (\$)	2700	2822	2948	3081	3220	3365

$FV = 2700(1 + 0.045)^n$

b The graph will have 'Year' on the horizontal axis and 'Value of investment' on the vertical axis. Plot the points and join them with a smooth curve.

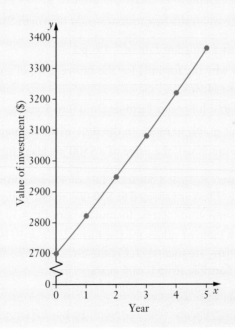

Exercise 6.03 Exponential growth and decay in finance

1 Mirabella invested $800 for 6 years at 9% p.a. compounding annually.

    a Complete this table of values for the value of the investment over 6 years.

Year	0	1	2	3	4	5	6
Value of investment ($)	800						

    b Draw a graph showing the growth of her investment over 6 years.

2 $A = 1000 \times (1.07)^n$ gives the value $A of $1000 invested at 7% p.a. compounding annually for n years.

    a Gemma invested $1000 at 7% p.a. compounding annually. How much will her investment be worth in 12 years?

    b Copy and complete the table below and use it to draw a graph for this formula from 0 to 12 years. Round answers to the nearest dollar.

Year	0	2	4	6	8	10	12
Value of investment ($)							

    c Use your graph to estimate the value of $1000 after 5.5 years.

    d Estimate when the investment would be worth $1600.

3 Inflation is at a constant rate of 2.5% p.a.

    a Explain why the formula for the future price of a necklace in n years is $A = 500 \times (1.025)^n$.

    b Draw a graph for this formula from 0 to 5 years.

    c Use your graph to estimate the time when the necklace will have a value of $545.

    d By how much will the price of a $500 necklace increase in the next 3 years?

4 Prices are increasing at a constant rate of 4% p.a. Imran is planning to buy a new car in 3 years time. The car is priced at $30 000 today. We can model the future price of his car using our compound interest formula where $PV = 30\ 000$ and $r = 0.04$.

    a This is a table of values for the formula $FV = 30\ 000(1.04)^n$. Two of the values in the bottom row are incorrect. Fix the errors and use the correct table to draw the graph for this formula from 0 to 5 years.

Number of years (n)	0	1	2	3	4	5
Price of car (P)	30 000	31 200	32 484	33 746	35 296	36 500

    b How much will the car cost at the end of 3 years?

    c Use your graph to estimate when the price of the car will be more than $34 000.

    d Mathematical models are based on assumptions and are only as good as the assumptions. Is it likely that prices will continue to increase at 4% p.a. for many years? What factors might affect the accuracy of this model of price increases?

ISBN 9780170413534

5 Kate's diamond necklace is appreciating at 12% p.a. This year, a jeweller has valued it at $4500.

   a Explain why the appreciated value can be calculated with the formula $FV = 4500(1 + 0.12)^n$.

   b Draw a graph of the changing value of Kate's necklace over the next 6 years.

   c Use your graph to estimate the value of Kate's necklace after 2.5 years.

6 Xavier bought a car priced at $30 000. The car depreciates at a rate of 15% p.a.

   a Explain why the salvage value can be calculated with the formula $S = 30\,000(0.85)^n$.

   b Use this formula to complete this table of the salvage value of the car over 6 years.

Year (*n*)	0	1	2	3	4	5	6
Salvage value ($*S*)	30 000						

   c Draw the graph for this table of values.

   d Use your graph to estimate how long it takes for the car to halve in value (to be worth $15 000).

7 Jennifer buys a computer workstation costing $7700 for her business. She is able to depreciate the workstation at 20% p.a. for tax purposes.

   a In the formula for declining-balance depreciation, $S = V_0(1 - r)^n$, what values will you use for V_0 and r in this question?

   b Copy and complete this table.

Year (*n*)	0	1	2	3	4	5
Salvage value ($*S*)	7700					

   c Draw the graph for this table of values.

   d Use your graph to estimate how long it takes for the computer workstation to halve in value (to be worth $3850).

   e Jennifer received some advice that she should replace her workstation when its value falls below $2500. Approximately when should Jennifer replace her workstation?

Graphing
exponentials

Practising
growth problems

6.04 Exponential modelling

Exponential models aren't just for investments and values of items. We can also model changes in populations of people, animals and bacteria with exponential models.

EXAMPLE 4

When a chef sneezed into a pot of soup, 80 bacteria entered the soup. The number of bacteria increased by 50% every hour.

a Complete a table of values to show the number of bacteria in the soup over the next 5 hours.

b Draw a graph showing the number of bacteria in the soup.

c How long does it take for the number of bacteria to double?

d The equation for calculating the number of bacteria, B, after t hours is given by $B = 80 \times (1.5)^t$. Calculate the number of bacteria after 12 hours.

ISBN 9780170413534

Solution

a   Number of bacteria at start = 80.
Every hour, number of bacteria
increases by 50%.
Find 50% and add
OR multiply by 150% or 1.5.
After the 1st hour,
bacteria = 80 + 40 = 120

After the 2nd hour,
bacteria = 120 + 60 = 180
and so on.

Time in hours	0	1	2	3	4	5
Number of bacteria	80	120	180	270	405	608

b   The graph will have 'Time in hours'
on the horizontal axis and 'Number
of bacteria' on the vertical axis.

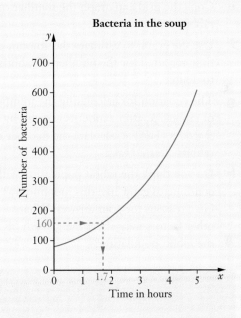

Bacteria in the soup

c   Find the time it takes the bacteria
to double, say, from 80 to 160. Read
from 160 on the vertical scale
(see graph).

The doubling time is approximately
1.7 hours.

d   Substitute $t = 12$ into the equation.

$B = 80 \times (1.5)^{12}$
$= 10\,379.70703 \ldots$
$\approx 10\,380$

Example
4

Exercise 6.04 Exponential modelling

If necessary, round answers to the nearest whole number.

1 The rabbit population on an island is out of control. Every month, the number of rabbits increases by 25%. On 1 January, there were 60 rabbits on the island.

a By how many did the number of rabbits increase during January?

b How many rabbits were on the island on 1 February?

c Determine the number of rabbits on the island on 1 March.

d Copy and complete this table of values.

Date	1 Jan	1 Feb	1 March	1 April	1 May	1 June
Number of rabbits	60					

e Copy the grid and graph the number of rabbits on it.

f Use the graph to estimate the amount of time required for the number of rabbits to double.

g The equation for calculating the number of rabbits, R, after t months is given by $R = 60 \times (1.25)^t$. Calculate the number of rabbits after 2 years.

h Do you think the rabbit population will continue to grow like this indefinitely? Give reasons for your answer. What other factors may affect the growth of the rabbit population?

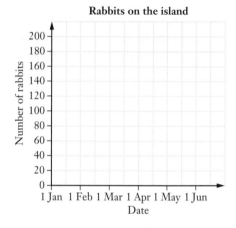

Rabbits on the island

ISBN 9780170413534

2 The world's population in 7000 BCE was estimated to be 2.4 million. Archaeologists believe that before the Common Era (BCE), the world's population was doubling every 1000 years.

a Copy and complete the table of values to show the world population from 7000 BCE to the year 0 (actually 1 CE).

Year BCE	7000	6000	5000	4000	3000	2000	1000	0
Population (millions)								

b Construct a graph showing the world population from 7000 BCE to the year 0.

c In 1999, the world's population reached 12 billion, or 12 000 000 000. Do you think that the world population doubling time after the year 0 was 1000 years? Give a reason to justify your answer.

3 A new strain of the flu is spreading through the community and the number of people with the flu is increasing by 6% per day. At the moment, 40 people in the community are infected. The equation for calculating the number of infected people, I, after t days is given by $I = 40 \times (1.06)^t$.

a Copy and complete this table showing the increasing number of people with the flu over 10 days. Remember to round your answers to the nearest whole person!

Number of days (t)	0	1	2	3	4	5	6	7	8	9	10
Number of infected people (I)	40										72

b Draw the graph from this table of values.

c Use your graph to estimate the time it will take to infect 55 people with the flu.

d Do you think the number of people with the flu can continue to increase exponentially? Give a reason for your answer.

4 This table of values shows the size of a population every 10 years.

Number of years	0	10	20	30	40	50
Population	60	72	86	104	124	149

a By how many did the population increase in the first 10 years?

b Calculate the percentage increase between year 0 and year 10.

$$\text{Percentage increase} = \frac{\text{number increased}}{\text{previous population}} \times 100$$

c Is the percentage increase the same for each 10-year period?

d Draw the graph for the information in the table.

e The formula for this table is approximately $P = 60 \times (1.2)^x$, where x is the number of 10-year periods. Use the formula to estimate the population after 100 years.

5 A day-old, baby grey kangaroo weighs 13 g and his weight is increasing by 7% per day. The equation that models this exponential growth is $W = 13 \times (1.07)^t$.

a Copy and complete this table of values to graph the increasing weight of the baby kangaroo over 12 days.

Day, t	0	1	2	5	7	10	12
Weight, W (g)							

b Estimate the kangaroo's age when his weight is 26 g.

c Do you think the kangaroo's weight will continue to increase exponentially? Justify your answer.

6 The number of people who have developed resistance to a new virus is doubling every year. In the year 2017, 50 people had resistance to the virus. The equation for this model is $P = 50 \times 2^n$.

a Copy and complete the table of values.

Number of years after 2017 (n)	0	1	2	3	4	5
Number of people with resistance (P)	50					

b Copy the grid and graph the number of people who have developed resistance to the virus over time.

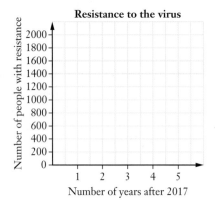

Resistance to the virus

c Use the formula to calculate the number of people who are resistant to the virus in 2037.

INVESTIGATION

HOW MANY FERAL CAMELS?

Thousands of camels were brought into Australia between 1840 and 1907 to be used for transport in dry areas of Central and Western Australia. Experts estimate that numbers of feral camels are now increasing at 10% per year and in 2018 there were 500 000 feral camels in Australia (after mass culling in 2013), which is having a devastating effect on the environment.

For this investigation, ask your teacher to download from NelsonNet the **Feral camels** spreadsheet. Work in small groups.

1 a Without making any calculations, predict the number of feral camels there will be in 2023 if there were 500 000 feral camels in 2018.

b Describe ways that the feral camels could be destroying the environment. Remember to think about both plants and animals.

2 Copy and complete this table using the spreadsheet, showing the 10% increase each year and the total number of feral camels at the end of 5 years.

	Number of camels
Start of 2018	500 000
Increase during 2018	
Number at the end of 2018	
Increase during 2019	
Number at the end of 2019	
Increase during 2020	
Number at the end of 2020	
Increase during 2021	
Number at the end of 2021	
Increase during 2022	
Number at the end of 2022	

3 How accurate was your prediction? Were you surprised by the number?

4 Why do you think the increase in the number of camels isn't the same every year?

5 Do you think camel numbers will increase in the same way indefinitely? Justify your answer.

6 Suggest some factors that could complicate the growth of camel numbers.

7 How many years will it take camel numbers to reach 2 million at:

a 10% p.a. growth? **b** 5% p.a. growth? **c** 7% p.a. growth?

8 Predict the annual growth rate that will make feral camel numbers greater than 2 million in 5 years, then use the spreadsheet to check your prediction.

9 Describe the similarities between the growth in camel numbers and how an investment grows with compound interest. Are there any differences?

6.05 Graphing practical situations

Some situations give graphs that don't follow an equation.

Graphing
practical
situations

EXAMPLE 5

Sarah is pouring water into this container at a constant rate.
Draw a graph showing the height of the water in the
container over time.

Solution

At first, the height of the water will increase at a steady rate.

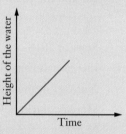

When the water reaches the narrower
section of the container, the height of
the water will increase more quickly, at a
constant faster rate.

EXAMPLE 6

This graph shows the amount of petrol in the tank of Natasa's
car over a week, starting on Monday. Describe what the
graph is showing.

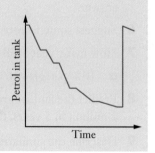

Solution

The amount of petrol goes down over a number of days. It then increases quickly before decreasing a small amount.	Natasa starts the week with a full petrol tank. As she drives throughout the week, the amount of petrol in the tank goes down.
Other descriptions are possible.	When the graph is horizontal (flat), Natasa isn't using the car.
	Towards the end of the week, Natasa again fills the tank and drives again.

Exercise 6.05 Graphing practical situations

1 Water is poured into each container at a steady rate. Match the container to the graph of the height of the water in the container.

a 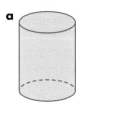   **b** 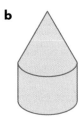   **c**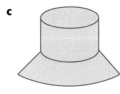

A 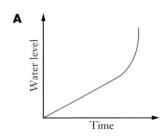   **B** 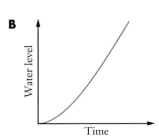   **C**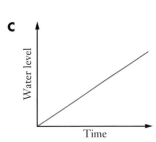

2 Filomena poured orange juice into this jug before her party. Draw a graph showing the height of the orange juice in the jug over time if she pours the juice at a constant rate.

3 A grain silo at a train siding is emptied at a constant rate into a goods train to be transported to the city. Draw a graph showing the height of the grain in the silo over time.

4 This graph shows the speed of a train as it travels from Strathfield to Lidcombe. Write a description of what happens.

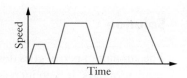

5 This graph shows the number of students at Nelson Valley High over 1 day. Write a description of how the number of students changes over the day.

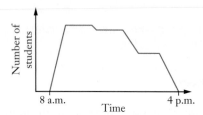

6 The graph below shows the temperature in Canberra over one day in July from midnight. Write a story to describe the change in temperature based on this graph.

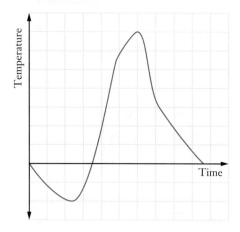

7 Draw a graph similar to that in Question **6** to show the change in temperature in your town or city over one day.

8 The incomplete graph below represents Hong's trip from Sydney to Goulburn and back.

   a Describe the section of the journey shown on the graph.

   b Hong stays in Goulburn for 2 hours and then returns to Sydney without stopping, arriving at 5 p.m. Copy the graph, extending the Time axis to 5 p.m., and complete the graph to show the return section of Hong's trip.

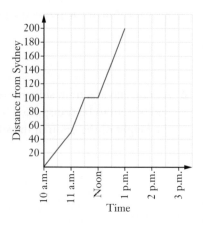

9 A lift is going up an 8-storey building. At ground level, 5 people enter the lift. 2 people get off on Level 2. One person gets off on Level 5. Two people get off on Level 8 where 3 other people get on the lift. They travel back to ground level.

Construct a graph to represent the height of the lift. Use time on the horizontal axis and building level on the vertical axis.

KEYWORD ACTIVITY

IN YOUR OWN WORDS

1 Describe in your own words what is meant by the term 'non-linear graph'.

2 Draw the shape of the curve that is associated with each term.

   a quadratic function

   b exponential growth

   c exponential decay

3 Write a definition of each term.

   a appreciation

   b compound interest

   c declining-balance depreciation

   d exponential decay

   e exponential growth

   f inflation

4 Choose one type of graph that you have used in this chapter and describe it.

SOLUTION TO THE CHAPTER PROBLEM

Problem

The number of people living on New Century Road is increasing by 9% p.a. and currently there are 90 people living there.

a Complete a table to show its population in the first 5 years. In your calculations, round to the nearest whole number.

b Draw a graph showing this population growth.

c The equation for calculating the population, P, after t years is given by $P = 90 \times (1.09)^t$. Calculate the population of New Century Road in 10 years.

d Suggest factors that could complicate the modelling of this real-life situation.

Solution

a

Number of years	0	1	2	3	4	5
Population	90	98	107	117	128	140

b

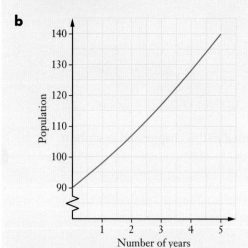

c $P = 90 \times (1.09)^t$

$= 90 \times (1.09)^{10}$

$= 213.062 \ldots$

$= 213$

The population of New Century Road in 10 years is 213.

d The calculation assumes that the rate of population increase is constantly 9% p.a. If the area becomes popular, a lot more people could move to the road as more homes are built. Also, if the area attracts young couples who are about to start families, the growth rate could increase.

Alternatively, if economic conditions change, people may not be able to afford to move here and the growth rate could fall.

- Give an example of a context that could be modelled using what you have learned in this chapter.
- Give an example of the formula related to the example you chose. If there is no formula, explain why.
- Is there any part of the topic you didn't understand? If so, ask your teacher for help.

Copy and complete this mind map of the topic, adding detail to its branches and using pictures, symbols and colour where needed. Ask your teacher to check your work.

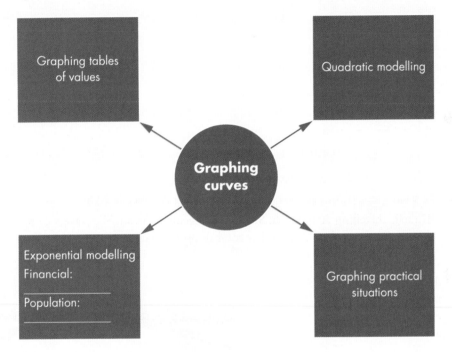

Exercise 6.01

1 Susie planted a new shrub in her garden. She measured its height each week at the same time. This table gives her data on the plant's growth.

Week	1	2	3	4	5	6
Height (cm)	3	7	10.5	13.5	16	18

Draw the graph of this data and join the points with a smooth curve.

Exercise 6.02

2 This graph shows the braking distance for a car at different speeds.

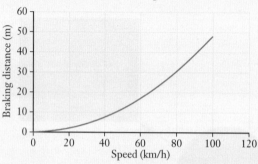

a What is the braking distance at 100 km/h?

b At what speed is the car travelling if the braking distance is 30 m?

c Bridie is travelling at 60 km/h. She sees an accident ahead. She travels 25 m before she applies the brakes. What is her total stopping distance?

Exercise 6.02

3 A bridge over a highway underpass has a shape modelled by the equation $h = 50 - 0.02x^2$, where h is the height above the road and x is the horizontal distance from the centre of the road.

Shutterstock.com/MNBB Studio

a Copy and complete this table.

Horizontal distance from centre of road, x (m)	−50	−40	−30	−20	−10	0	10	20	30	40	50
Height above road, h (m)											

b Graph the equation.

c What is the height of the centre of the bridge?

d Estimate the values of x where the height of the bridge above the road is half of the height of the bridge at the centre of the road.

4 Ali invested $9000 for 5 years at 4% p.a. compounding annually.

Exercise 6.03

  a Copy and complete this table for the value of his investment over 5 years.

Year	0	1	2	3	4	5
Value of investment ($)	9000					

  b Draw a graph showing the value of Ali's investment over 5 years.

  c From your graph, estimate when Ali's investment is worth $9500.

5 Iris buys a new car priced at $47 000. The car depreciates at a rate of 15% p.a.

Exercise 6.03

  a In the formula for declining-balance depreciation, $S = V_0(1 - r)^n$, what values will you use for V_0 and r in this question?

  b Copy and complete this table of the salvage value of the car over 6 years.

Year	0	1	2	3	4	5	6
Salvage value ($)	47 000						

  c Draw the graph for this table of values.

  d Use your graph to estimate how long it takes for the car to halve in value (to be worth $23 500).

6 The number of bacteria in a Petri dish over time is given in the table below.

Exercise 6.04

Day	0	1	2	3
Number of bacteria	2000	4000	8000	16 000

  a Draw the graph from this table of values.

  b Use your graph to estimate when there will be 5000 bacteria in the Petri dish.

  c The equation that models this bacteria growth is $B = 5000 \times 2^n$, where n is the number of days. How many bacteria will there be after 10 days?

7 Draw a graph that represents this situation.

Exercise 6.05

The noise level in a classroom varies over the one hour of a lesson. The class is quite noisy as they enter the classroom. They then become silent to listen to the teacher explain what is going to happen in the lesson. Some students ask questions before the class is divided into groups to work together for the rest of the lesson. Group work is very noisy. Five minutes before the end of the lesson the teacher asks for silence so she can finish the lesson.

8 The graph on the right shows the journeys of 2 cyclists on a Sunday bike ride. Describe the journeys of both cyclists.

Exercise 6.05

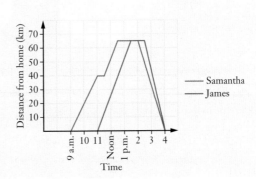

7.

APPLYING RATES

Chapter problem

Suzie needs to buy some concrete sealer to seal the bricks in her new garden wall. She needs 24 litres and the sealer is available in 4 L and 6 L containers.

$55

concrete sealer

4 L

concrete sealer

$87

6 L

What is the cheapest way for Suzie to buy the sealer?

CHAPTER OUTLINE

WHAT WILL WE DO IN THIS CHAPTER?

- Simplify rates
- Convert rates from one pair of units to another, for example, m/s to km/h
- Solve practical problems using rates, including speed, fuel consumption and unit pricing
- Interpret distance-time (travel) graphs

HOW ARE WE EVER GOING TO USE THIS?

- Comparing prices of different brands of groceries to select the cheapest
- Calculating how long a trip will take if driving at a certain speed
- Determining quantities required for a painting or tiling job
- Calculating the quantity of fuel a car will use on a trip

7.01 Rates

While ratios compare two quantities of the same type, **rates** compare two different types of quantities.

The most common rate we use is **speed,** which compares distance travelled over time. If a car has a speed of 100 km/h, it means that the car travels 100 km for *each* hour it travels.

Unlike ratios, rates have units. We express these as the units of the first quantity **per** unit of the second quantity. 'Per' means 'for each one'.

EXAMPLE 1

Write each situation as a rate.

a Andrew runs 200 m in 25 seconds.

b The fuel for Boris' cultivator costs $35 for 20 L.

Solution

a The units for this rate will be metres per second, or m/s.
Divide the number of metres by the number of seconds.

$$\text{Rate} = \frac{200 \text{ metres}}{25 \text{ seconds}}$$
$$= 8 \text{ m/s}$$

b The units will be dollars per litre.
Divide the number of dollars by the number of litres.

$$\text{Rate} = \frac{\$35}{20 \text{ litres}}$$
$$= \$1.75/\text{L}$$

Exercise 7.01 Rates

Example 1

1 Write each situation as a rate.

    a 200 km travelled in 4 hours

    b 75 words typed in 3 minutes

    c 250 L used in 5 hours

    d $87.50 for 5 kilograms

    e 400 m in 55 seconds

    f $65 earned in $2\frac{1}{2}$ hours

    g 500 g in 2 litres

    h 1800 revolutions in 3 minutes

2 Marcella uses 90 litres of composting manure on her flower bed which has an area of 5 m^2. What is this as a rate in L/m^2?

3 Goran used his sprinklers to water his garden for $2\frac{1}{2}$ hours. He used 3375 litres of water. What is his rate of water use in L/h?

ISBN 9780170413534

4 Write each situation as a rate.

a 320 L in 8 containers

b 10 litres consumed in $2\frac{1}{2}$ days

c 80 g in 15 cm^3

d $22.40 to hire 7 DVDs

e 175 mm rain in 5 days

f 290 km driven using 20 L of petrol

g $20.52 for 1.8 m of curtain material

h 768 vibrations in 3 seconds

i 140 sheep on 3.5 hectares of land

j $28.44 for 18 L of petrol

5 Dakota used 8 L of stone sealer to seal 72 m^2 of slate. Calculate this rate in m^2/L.

6 John is preparing his budget for his holiday. For his 10-day holiday he is budgeting $1200 for meals. Calculate John's meal budget in $/day.

7 Marcus runs a sausage sizzle for his sports club's market day. He has purchased the food shown in the table. Calculate the cost rate, including units, for each item.

Item	Cost and amount	Cost rate
Sausages	$57.80 for 6.8 kg	a
Steak	$93.60 for 7.5 kg	b
Onions	$9.80 for 4 kg	c
Rolls	$132 for 240 rolls	d
Tomato sauce	$19.60 for 8 bottles	e

7.02 Distance, speed and time

Problems involving distance, speed and time are very common, and involve applications of rates. When we know 2 of the values we can calculate the third value.

Speed formula

$$\text{Distance covered} = \text{speed} \times \text{time}$$
$$\text{or } D = S \times T$$

The units for speed tell us the units for distance and time.
When the speed is in km/h, the distance is in kilometres and the time is in hours.

The distance, speed and time triangle

Some students find it easier to use the 'distance, speed and time triangle' to solve problems involving speed. Place the letters D for distance, S for speed and T for time in alphabetical order in a triangle.

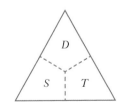

To calculate speed, cover up S, which leaves $\dfrac{D}{T}$.

This means that $S = \dfrac{D}{T}$, or $S = D \div T$.

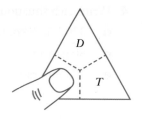

The formulas for time (T) and distance (D) can be found in the same way.

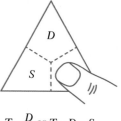

$$T = \frac{D}{S} \text{ or } T = D \div S$$

$$D = S \times T$$

EXAMPLE 2

A racing greyhound runs at a speed of 18 m/s.

a How far will the greyhound run in 4 seconds?

b How long will it take the greyhound to complete a 1200 m race?

Solution

a Because we're calculating distance, cover the D, leaving S and T together. To calculate D, we multiply S and T.

$D = S \times T$
$S = 18$ m/s and $T = 4$ s
$D = 18 \times 4$
$\quad = 72$ m

Write your answer.

The greyhound will run 72 m in 4 s.

b Because we're calculating time, cover the T, which leaves D over S, or $D \div S$.

$T = D \div S$
$D = 1200$ m and $S = 18$ m/s
$T = 1200 \div 18$
$\quad = 66.7$ seconds

Write your answer.

The greyhound will complete the race in 66.7 s.

EXAMPLE 3

A kangaroo bounds at a speed of 48 km/h. How far will a kangaroo bound in 20 minutes?

Solution

The formula is $D = S \times T$
The speed is in km/h, so we need time in hours.
Divide 20 minutes by 60 to change to hours.

$$S = 48 \text{ km/h}$$
$$T = \frac{20}{60} \text{ hour}$$
$$= \frac{1}{3} \text{ h}$$
$$D = S \times T$$
$$= 48 \times \frac{1}{3}$$
$$= 16 \text{ km}$$

Write your answer.

The kangaroo will bound 16 km.

Distance-time graphs

Distance-time graphs

A **distance-time graph** is a line graph that describes a journey, by comparing distance (on the vertical axis) with time (on the horizontal axis). The slope or steepness of the graph indicates speed.

EXAMPLE 4

This graph shows Monique's cycling trip.

a   At what time did Monique leave home?

b   When was Monique's first stop? How far from home was she?

c   Find her average speed over the first 2 hours.

d   What time was it when Monique began her journey home?

e   How far did she travel all together?

f   Find her average speed during the trip home.

g   For how long did Monique stop altogether during the trip?

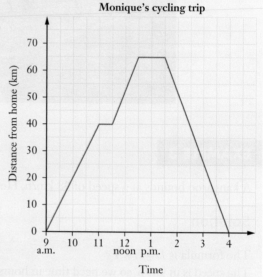

Monique's cycling trip

Solution

a   Monique left home at the start of the graph, when the distance was 0.

Monique left home at 9 a.m.

b   Monique first stopped where the graph is flat. The distance from home does not change, which means Monique has stopped.

Monique first stopped at 11 a.m., 40 km from home.

c   Distance = 40 km, time = 2 h.
$$S = \frac{D}{T}$$

$$\text{Average speed} = \frac{40\,km}{2\,h}$$
$$= 20 \text{ km/h}$$

d   Monique started returning home where the graph points downward.

Monique started returning home at 1:30 p.m.

e   Monique travelled 65 km, then returned home.

Total distance = 2×65 km
$$= 130 \text{ km}$$

f Distance = 65 km, time = $2\frac{1}{2}$ h

$$S = \frac{D}{T}$$

Average speed $= \dfrac{65\,\text{km}}{2\frac{1}{2}\text{h}}$

$= 26$ km/h

g Look at the places where the graph is flat.

First stop: $\dfrac{1}{2}$ hour.

Second stop: 1 hour

Total stopping time $= \dfrac{1}{2} + 1 = 1\dfrac{1}{2}$ hours

On a distance-time graph:

- a horizontal (flat) section on the graph indicates a stop
- the steeper the line, the greater the speed (more distance covered in less time)
- a section going down, towards the right, indicates a change in direction or that the traveller is returning towards the start.

Exercise 7.02 Distance, speed and time

1 a Use the formula $T = \dfrac{D}{S}$ to determine the value of T when $D = 80$ and $S = 16$.

b In the formula $D = S \times T$, what is the value of D when $S = 80$ and $T = 3$?

2 a Karen is driving at a speed of 60 km/h. How far will she drive in 3 hours?

b How long will it take her to drive 240 km?

Example 2

3 Brian has an appointment 160 km away. He must be there in 2 hours. At what average speed must he travel to arrive in time?

4 Go-karts race at a speed of 110 km/h. At this speed, how many kilometres can a go-kart travel in a $2\frac{1}{2}$ hour race?

Shutterstock.com/Jaggat Rashidi

5 Wasim is driving through heavy traffic at a speed of 32 km/h. How far will he travel in 15 minutes?

6 An ambulance is racing to the scene of a serious accident at a speed of 100 km/h. The accident is 15 km from the ambulance station.

   a How long will the ambulance take to reach the accident? Express your answer as a decimal of an hour.

   b Multiply your answer to part **a** by 60 to change the time to minutes.

7 A crisis team is travelling at 80 km/h to assist in a siege situation 10 km away. How long will it take the team to arrive at the scene? Express your answer in minutes.

8 A whitewater rafting team completed three sets of rapids and 1.6 km of calm water in 30 minutes. The lengths of the sets of rapids were 150 m, 80 m and 170 m.

   a Calculate the distance that the rafting team covered in kilometres.

   b Explain why you can't use $T = 30$ in the equation $S = D \div T$ to calculate the speed of the raft in km/h.

   c Calculate the raft's average speed in km/h.

9 Cheetahs can run at a speed of 31 m/s in short bursts.

  a   How far can a cheetah run in 9 seconds?

  b   How long does it take a cheetah to run 140 m, correct to one decimal place?

10 A peregrine falcon's top speed is 90 m/s.

  a   How far can a falcon fly in 1 minute?

  b   How long will it take the falcon to fly 1 km?

11 The table shows distances between several cities in kilometres.

Approximate kilometres	Albury	Brisbane	Canberra	Goulburn	Sydney	Tamworth
Albury	–	1610	190	380	600	1040
Brisbane	1610	–	1300	1225	1020	575
Canberra	190	1300	–	95	300	750
Goulburn	380	1225	95	–	205	660
Sydney	600	1020	300	205	–	460
Tamworth	1040	575	750	660	460	–

  a   How far is it from Canberra to Tamworth?

  b   How long will it take to drive from Canberra to Tamworth at an average speed of 75 km/h?

  c   Gordana took 5 hours to drive from Goulburn to Albury. What was her average speed?

  d   Max and Wayne left Albury at 6 a.m. on Monday to drive to Brisbane. They shared the driving and completed the trip at an average speed of 70 km/h. At what time did they arrive in Brisbane?

Example 4

12 Kyle and Anna went to the movies together. The graph shows the distance they were from Kyle's house after the movie finished.

a How far is the cinema from Kyle's house?

b After the movie finished, they rode their bikes to Anna's house. How far is Anna's house from the cinema?

c How long did it take them to ride from the cinema to Anna's house?

d What was their average speed riding to Anna's house?

e How long did Kyle stay at Anna's house?

f Calculate Kyle's average speed when he was riding home.

g Part of the trip home is downhill and part is uphill. Did Kyle ride the uphill section during the first or second hour of his ride home? Justify your answer.

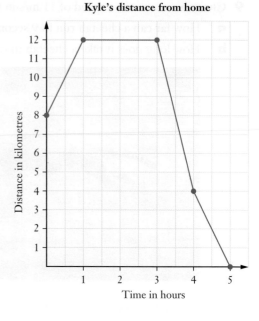

Kyle's distance from home

13 This graph shows a cyclist's day trip.

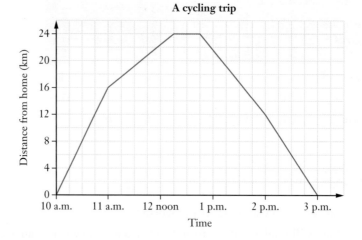

A cycling trip

a At what time did the speed of the cyclist

  i increase?   **ii** decrease?

b When did the cyclist start to return home?

c How far did the cyclist travel altogether on this day?

d How long did the cyclist spend 'on the road'?

e Find the cyclist's average speed for:

  i the first hour   **ii** 11:00 a.m. to 12:15 p.m.

  iii 12:45 p.m. to 2:00 p.m.   **iv** the entire day

14 Brian travels from Bligh to Macquarie, while Sam travels from Macquarie to Bligh.

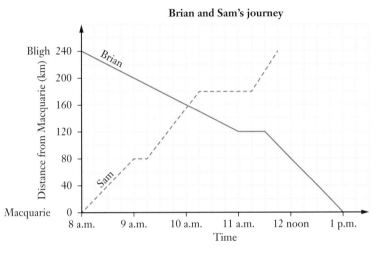

a How far is it between the 2 towns?

b Who is travelling faster? How can you tell?

c How far is Brian from Macquarie at 11 a.m.?

d How far is Sam from Macquarie at 11 a.m.?

e At what time do Brian and Sam pass each other? How far are they from Macquarie when they pass?

f Is Sam travelling faster before 9 a.m. or after 9:15 a.m.? How does the graph show this?

g Calculate Brian's average speed before he stops.

h For how long did Sam stop altogether on the trip? Select the correct answer **A**, **B**, **C** or **D**.

  A 60 minutes   **B** 75 minutes   **C** 5 minutes   **D** 90 minutes

15 Use the clues to determine which person matches with each graph. Copy and complete table below to record your answers.

Clues

- Shelby walks home at a speed of 4 km/h.

- Walter rides his bike at an average speed of 12 km/h.

- Peta travels on the slow bus. With all the stops, the bus averages only 8 km/h.

- Luke drives home at an average speed of 60 km/h.

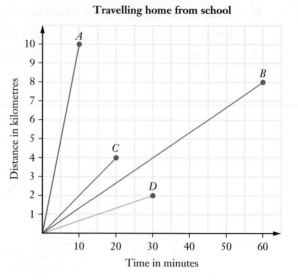

Travelling home from school

Graph	Person	Distance from school
A		
B		
C		
D		

16 For yourself and 3 other people in your class, construct a graph similar to the graph in Question **15** that shows your trips home from school.

SPEED UNITS

In this activity, you are going to investigate a variety of speeds using the Internet. In particular, you are going to look at the units used to express speed. Then, you are going to develop a rule for unit selection to help you answer the question: 'What unit should I use?'

Copy the following table to help you record your results. Alternatively, print a copy from NelsonNet.

Speed units

Types of speeds	Typical or record speed
	Some activities may have speeds expressed in more than one way; for example, m/s and km/h. Record all unit types.
Slow speeds	
Moving snail	
Adult walking	
Athlete running	
Medium speeds	
Cruise ship	
Car	
Downhill, cross-country skiing	
Olympic speed skating	
Fast speeds	
Queensland tilt train	
Japanese high-speed train	
International airline jet	
Very fast speeds	
Sound Bullet fired from an assault weapon Fighter jet International Space Station Light	

Group discussion questions

1 In what situations do we use the following speed units?

m/h   m/s   km/s   km/h

2 Why might we decide to measure the same speed using different speed units?

Converting
rates

More rates

7.03 Converting rates

Sometimes, it is useful to be able to convert rates from one set of units to another. Doing this helps us to compare rates for different circumstances.

Converting rates from one set of units to another

- Write the rate as a fraction, including its units
- Convert the units on the numerator of the fraction to the new unit
- Convert the units on the denominator of the fraction to the new unit
- Simplify the rate.

Converting
rates

EXAMPLE 5

Convert 90 km/h to metres/second (m/s).

Solution

Convert 90 km into metres and 1 hour into seconds.

$$90 \text{ km} = 90 \times 1000 \text{ metres}$$
$$= 90\,000 \text{ m}$$
$$1 \text{ h} = 60 \text{ minutes}$$
$$= 60 \times 60 \text{ seconds}$$
$$= 3600 \text{ s}$$

Divide the number of metres by the number of seconds.

$$90 \text{ km/h} = \frac{90 \text{km}}{1h}$$
$$= \frac{90000 \text{m}}{3600 \text{s}}$$

Simplify the rate.

$$= 25 \text{ m/s}$$

NCM 12. Mathematics Standard 1

ISBN 9780170413534

Convert 2.5 g/m² to kg/ha.

> There are 10 000 m² in 1 hectare.

Solution

Change 2.5 g to kg and 1 m² to hectares.

$$2.5 \text{ g} = 2.5 \div 1000 \text{ kg}$$
$$= 0.0025 \text{ kg}$$
$$1 \text{ m}^2 = 1 \div 10\,000$$
$$= 0.0001 \text{ ha}$$

Simplify the rate.

$$2.5 \text{ g/m}^2 = \frac{2.5\text{g}}{1\text{m}^2}$$
$$= \frac{0.0025 \text{ kg}}{0.0001 \text{ ha}}$$
$$= 25 \text{ kg/ha}$$

Exercise 7.03 Converting rates

In this exercise, round answers to one decimal place where needed.

1 Convert each speed into m/s.

  a   20 km/h   **b**   50 km/h   **c**   80 km/h   **d**   110 km/h

2 Alice is walking at a rate of 80 metres per minute. What is her speed in:

  a   km/h?   **b**   m/s?

3 Convert each speed to km/h.

  a   12.5 m/s   **b**   2400 m/minute   **c**   3500 m/h

4 Joshua fertilises his land at the rate of 4 g/m². Express this rate in kg/hectare.

5 A tap leaks at a rate of 30 mL per minute. What is this rate in litres per hour?

6 The steel mesh used in fencing weighs 1.2 kg/m.

  a   Convert this rate to g/mm.

  b   Convert this rate to tonnes/km.

  c   What do you notice about the answers to parts **a** and **b**? Can you explain why?

  d   What is the rate in g/cm?

7 Jesse pumps water from his dam to irrigate his market gardens. The pump delivers water at a rate of 3.5 kL/hour. Express this rate as:

   a   L/h   **b**   L/min   **c**   L/s.

8 A 5 L container of varnish costs $14.00.

   a   What is the cost in $/L?

   b   Express this as a cost in cents/mL.

9 Jamaica's Usain Bolt set the world record for the 100 metre sprint. He ran the distance in 9.58 seconds.

   a   Express this as a rate in m/s (to one decimal place).

   b   Convert this speed to km/h (to two decimal places).

Getty Images/AFP

7.04 Unit pricing

The government requires supermarkets to display the **unit price** for all items. The unit price makes comparing the prices of different-sized packaging easy.

Best buys puzzle

Alamy Stock Photo/Jeff Greenberg

Jonathan is calculating the unit price for rolls of paper towels. He has 15 m rolls for $1.65 each and 20 m rolls at $2.40 per roll. He decided to use 1 metre as the comparison unit.

a   Calculate the price per metre for each roll.

b   Which roll is the better value for money?

Solution

a   For the 15 m roll, calculate the cost of 1 metre by dividing the cost of the roll by 15.

For the 20 m roll, calculate the cost of 1 metre by dividing the cost of the roll by 20.

For the 15 m roll, cost of 1 m
$= \$1.65 \div 15$
$= \$0.11$

For the 20 m roll, cost of 1 m
$= \$2.40 \div 20$
$= \$0.12$

b   The better price is the lower price for 1 m.

The 15 m roll is the better value.

Unit pricing

EXAMPLE 8

The same cooking oil is available in two sizes of bottle:

500 mL for $4.75 and 750 mL for $6.45.

> A 'unit' varies depending on the items being compared. It can be any convenient unit; for example, 100 g, 1 L, 100 mL or 1 m.

a Calculate the unit price for each bottle.

b Which size of bottle is the better value?

Solution

a For each bottle, calculate the price for 100 mL. 500 mL is 5 × 100 mL

Divide the price of the 500 mL bottle by 5.

Unit price of the 500 mL bottle is:
$4.75 ÷ 5 = 0.95
100 mL costs $0.95

750 mL is 7.5 × 100 mL

Divide the price of the 750 mL bottle by 7.5.

Unit price of the 750 mL bottle is:
$6.45 ÷ 7.5 = 0.86
100 mL costs $0.86

> We get 7.5 by dividing 750 mL by the unit size, 750 ÷ 100 = 7.5.

b The bottle with the smaller unit price is the better value.

The 750 mL bottle is the better value.

Exercise 7.04 Unit pricing

Example 7

1 Calculate the price of 1 m of wallpaper for each size roll.

a 5 m for $11.50

b 10 m for $22.00

c 20 m for $45.00

2 The same seafood sauce is available in 50 mL bottles for $3.60 and 80 mL bottles for $5.60.

a For each size of bottle, calculate the price per 10 mL.

b Which size of bottle is the better value?

3 A box containing 1 dozen eggs costs $4.80 and a box containing 6 eggs is $2.10.

a Calculate the price per egg in each box.

b Which size box is the better value?

c Damien is buying 18 eggs. What is the cheapest way for him to do it?

4 Supermarket signs show the packet price and the unit price for 3 packets of breakfast cereal. Which size packet is the best value for money? Justify your answer.

450 g for $4.49
Unit price $1.00 per 100 g

350 g for $3.00
Unit price $0.86 per 100 g

330 g for $4.25
Unit price $1.29 per 100 g

5 Rice is available in 3 packets:

- 1 kg for $3.50

- 2 kg for $8.00

- 5 kg for $17.99

  a What would be a sensible unit to use for a price comparison?

  b Determine which size packet is the best value.

6 The yellow spice, saffron, is the world's most expensive spice.

The price of 1 g of saffron is 3 times the price of 1 g of 24 carat gold.

Example **8**

Carmen is ordering some saffron online.

2 g for $290

2.5 g for $350

5 g for $695

7.5 g for $1065

a Calculate the unit (1 g) price for each size.

b Carmen is going to order 7.5 g. What is the cheapest way for her to do it?

7 Breakfast biscuits are on special because the 'use before date' is only 5 weeks away. A box containing 24 biscuits is priced at $4.56 and a box containing 48 biscuits is $8.64.

Mae likes to have one biscuit for breakfast on most days. Which size box do you recommend she buy?

8 Hair shampoo is available in two sizes of bottle:
- 750 mL for $8.95
- 1 L for $11.50

Which size bottle is the better value? Justify your answer.

7.05 Rate problems

Rate problems

Rate skills

To solve problems involving rates, we either **multiply** or **divide** by the rate:
- Identify the rate and express the units of the rate as a fraction (e.g. $\dfrac{\text{km}}{\text{h}}$).
- To find the numerator of the fraction, multiply by the rate.
- To find the denominator of the fraction, divide by the rate.

Rate problems

EXAMPLE 9

Noah hires a small car for $42 per day. How much will the hire cost be if Noah keeps the car for 16 days?

Solution

The rate is $42/day. We can express the units as $\dfrac{\$}{\text{day}}$.

Hire cost = days × rate
= 16 days × $42/day
= $672

We need to find the number of dollars, which is the numerator of the fraction, so we multiply by the rate.

Check that the answer is realistic.

It will cost Noah $672 to hire the car for 16 days.

EXAMPLE 10

Rate
problems

Amy sends the fruit she grows on her farm to market in cartons that hold 12 kg of fruit per carton. How many cartons will she need to pack 180 kg of fruit?

Solution

The rate is 12 kg/carton. The units are $\dfrac{\text{kg}}{\text{carton}}$.

We need to find the number of cartons, which is the denominator of the fraction, so we divide by the rate.

Check that the answer is realistic.

Number of cartons
= weight ÷ rate
= 180 kg ÷ 12 kg/carton
= 15 cartons

Amy will need 15 cartons.

Exercise 7.05 Rate problems

1 Lizzie plants tomato seedlings in rows at a rate of 28 seedlings per row. How many seedlings are there in:

a 4 rows?   **b** 7 rows?   **c** 12 rows?

Example 9

2 Jayden earns $12.40/h as a barista in a cafe. How much is he paid for working 20 hours?

3 Kane proudly claims that he can hike at an average rate of 5 km/h even in difficult conditions. At this rate, how long will it take him to hike:

a 15 km?   **b** $7\frac{1}{2}$ km?   **c** 24 km?

Example 10

4 Mandarins cost $3.20/kg. How many kilograms of mandarins can Peter buy for $20?

5 Danielle is growing nectarine trees. She can expect a yield of 80 kg of fruit per tree.

a How many kilograms of nectarines can Danielle expect to produce if she plants 9 trees?

b How many trees would Danielle need to plant if she wanted to produce 1200 kg of nectarines?

6 A plane flies at a speed of 840 km/h.

a How far does the plane travel in $6\frac{1}{2}$ hours?

b Calculate how long it would take the plane to fly from Melbourne to each city, given their distances. Answer in hours correct to one decimal place.

   i Brisbane: 1670 km   **ii** Auckland, New Zealand: 2626 km

   iii Perth: 3430 km   **iv** Los Angeles, USA: 12 773 km

7 The yields Maxine achieves from different vegetable crops are shown below.

Crop	Yield (vegetables/m^2)
Beetroot	40
Lettuce	18
Cabbage	8

a How many beetroots can Maxine expect to get from her 24 m^2 beetroot plot?

b How many lettuces will she grow in her 15 m^2 plot?

c What area does she need to grow 450 lettuces?

d What area does she need to grow 760 cabbages?

8 Aaron uses fertiliser on his farm at a rate of 24 kg/hectare. How many kilograms of fertiliser does he require to cover 32 hectares?

9 A dripping tap leaks water at the rate of 25 mL/minute.

a How much water will leak from the tap in 30 minutes?

b How much water will be lost in 1 hour?

> Remember, there are 60 minutes in an hour.

c Convert your answer to part **b** to litres.

d How many litres will be lost in 24 hours?

e How many minutes will it take for the tap to leak 1 litre of water?

> Remember, there are 1000 mL in 1 L.

f How many hours will it take for the tap to leak 60 L of water?

10 Vijay is saving to purchase a car. He can save $400 per week.

a How much can Vijay save in 20 weeks?

b The car he wants costs $14 000. How long will it take Vijay to save this amount?

c A newer car is available but it costs $15 200. How much longer would it take Vijay to save for this car?

11 Chloe earns $24.50/h as a secretary. How many hours will she need to work to earn:

a $441?   **b** $857.50?   **c** $588?

12 Elise wants to hang curtains on 12 windows in her house. Each window requires 1.8 metres of material, which costs $12.50 per metre.

a How many metres of material will Elise need for all 12 windows?

b How much will it cost?

c Elise has budgeted only $225 for the material. How many metres of material can she buy if she sticks to this budget?

d How many windows can she curtain in this situation?

          ISBN 9780170413534

13 Jacob uses a small aircraft to do crop dusting. His plane uses fuel at a rate of 28 litres per hour. In one week, Jacob flew his plane for 18 hours.

a How many litres of fuel did Jacob use that week?

b That week, aviation fuel cost $2.25/L. How much was his fuel bill for the week?

c Jacob charges $124/h when he is crop dusting. How much did he charge for the week crop dusting?

d In a different week, Jacob had only 420 litres of fuel available. How many hours could he fly in that week?

7.06 Fuel consumption

In Year 11 Chapter 13, Buying a car, we examined the fuel consumption of cars, another example of rates.

Fuel consumption

> **Fuel consumption** is a rate expressed as the number of litres used per 100 km travelled (L/100 km). The lower the rate, the less fuel the car uses.
>
> $$\text{Fuel consumption} = \frac{\text{fuel used in L}}{\text{distance travelled in km}} \times 100$$

How much petrol?

There are 3 types of fuel consumption questions.

• What is the fuel consumption rate in L/100 km? Divide the fuel used by the distance travelled and multiply by 100.

• How much fuel will the car use? To find litres (L), multiply by the rate.

• How far can the car go? To find km travelled, divide by the rate.

EXAMPLE 11

The dashboard computer in Sue's car shows that over the last 5000 km travelled the car used fuel at a rate of 6.9 L/100 km.

a How much fuel did Sue's car use in the last 5000 km?

b At an average price of $1.55/L, how much did the petrol to travel 5000 km cost?

Solution

a Express 6.9 L/100 km as $\dfrac{6.9 \text{ L}}{100 \text{km}}$

To find litres, multiply by the rate.

$\text{Fuel consumption} = \dfrac{6.9 \text{ L}}{100 \text{ km}} = 0.069 \text{ L/km}$

$\text{Fuel used} = 5000 \text{ km} \times 0.069 \text{ L/km}$
$\qquad\quad = 345 \text{ L}$

b Express $1.55/L as $\dfrac{\$1.55}{\text{L}}$

To find cost ($), multiply by the rate.

$\text{Cost} = 345 \text{ L} \times \$1.55/\text{L}$
$\qquad\, = \$534.75$

EXAMPLE 12

Jesse bought a second-hand car. This week, he travelled 425 km and used 32 L of fuel. Calculate correct to one decimal place the car's fuel consumption in L/100 km.

Solution

$$\text{Fuel consumption} = \frac{\text{fuel used in L}}{\text{distance travelled in km}} \times 100$$

$$\text{Fuel consumption} = 32 \text{ L} \div 425 \text{ km} \times 100$$
$$= 7.5294 \dots$$
$$\approx 7.5 \text{ L/100 km.}$$

EXAMPLE 13

Saskia's BMW has a fuel consumption rate of 7.1 L/100 km.
How far (to the nearest kilometre) can the car travel on 60 L of fuel?

Solution

Express 7.1 L/100 km as $\dfrac{7.1 \text{ L}}{100 \text{ km}}$
To find km, divide by the rate.

$$\text{Fuel consumption} = \frac{7.1 \text{ L}}{100 \text{ km}} = 0.071 \text{ L/km}$$
$$\text{Distance} = 60 \text{ L} \div 0.071 \text{ L/km}$$
$$= 845.0704 \dots$$
$$\approx 845 \text{ km}$$

Exercise 7.06 Fuel consumption

1 The petrol consumption rate of Jack's car is 5.9 L/100 km.
  a How many litres of petrol will the car use on a 850 km trip?
  b How much will the fuel cost for the trip if petrol costs $1.60/L?

2 Tia's old Holden used 25 L of fuel to travel 180 km. Calculate the car's fuel consumption in L/100 km.

3 Jeff's VW Golf uses diesel fuel. It has a fuel consumption rate of 6.1 L/100 km. How far can Jeff travel on 50 litres of diesel fuel?

4 Charles used 18 L of petrol to drive 190 km. Calculate his car's fuel consumption in L/100 km.

5 A motoring website tested the fuel consumption for a number of SUV vehicles by measuring the fuel consumed for the same 790 km journey. Calculate the fuel consumption of each vehicle correct to one decimal place.

	Vehicle	Litres of fuel used
a	Kia SUV	80.1
b	BMW	69.9
c	Toyota Rav4	67.2
d	Subaru Outback	57.7
e	Range Rover	61.6

6 Carol's second-hand Toyota Camry uses 7.5 L/100 km. The petrol tank holds 60 litres of fuel. How far can Carol travel on one tank of petrol?

7 To save fuel, the motor in Lisa's BMW SUV turns itself off when the car stops in traffic or at traffic lights. With this fuel-saving option turned on, the car uses 8.9 L/100 km in city traffic but with it turned off it uses 10.0 L/100 km. Each week, Lisa travels 650 km in city traffic.

   a How much less fuel per week does Lisa use with the fuel-saving option turned on?

   b At an average price of $1.50/L, how much does Lisa save by turning on the fuel-saving option?

8 Alistair's Lexus Sport uses premium fuel (98 petrol) with a fuel consumption of 11.1 litres per 100 kilometres.

   a How much petrol will his car use on a 425 km trip to the vineyards? Answer correct to the nearest litre.

   b Calculate the cost of fuel for the journey at $1.75 per litre.

9 Ben's chainsaw uses fuel depending on the hardness of the wood he is cutting. When cutting large-diameter gum trees, it uses fuel at a rate of 310 mL in 45 minutes.

   a Calculate this fuel usage in L/h. Answer correct to one decimal place.

   b Ben estimates that it will take him 8 hours to cut up a large gum tree. At this fuel usage rate, how much fuel will Ben require?

   c Ben only has 3 L of fuel. How long will he be able to use the chainsaw?

10 Josh lives on a large property. He uses a ride-on mower to cut the grass, but he's heard that ride-on mowers use much more fuel than a hand mower and he's concerned about the effect on the environment. He decided to investigate. He divided his biggest paddock into two equivalent sections and mowed one with a ride-on and the other with a hand mower. He summarised the results in a table.

Method	Fuel usage rate	Time taken
Hand mower	600 mL/h	3 hours
Ride-on mower	3.6 L/h	30 minutes

   a Calculate the actual amount of fuel Josh used to mow each section.

   b Which method of cutting the grass would you recommend Josh use? Why?

CLUELESS CROSSWORD

Copy this crossword and position the keywords from this chapter in the crossword.
Then, write a set of clues for your crossword.

comparison   convert   fuel consumption   per

rate   speed   time   unit pricing

ISBN 9780170413534

SOLUTION TO THE CHAPTER PROBLEM

Problem

Suzie needs to buy some concrete sealer to seal the bricks in her new garden wall. She needs 24 litres and the sealer is available in 4 L and 6 L containers.

What is the cheapest way for Suzie to buy the sealer?

Solution

There are two ways to solve this problem!

First method: Calculate the cost per litre

Cost per litre in the 4 L container = $55 ÷ 4

$$= \$13.75$$

Cost per litre in the 6 L container = $87 ÷ 6

$$= \$14.50$$

It is cheaper to buy the sealer in 4 litre containers, so 6×4 L containers.

Second method: Calculate the total price of the containers Suzie would have to buy

Suzie needs 24 L. If she buys 4 L containers she will need 6. If she buys 6 L containers she will need 4.

Total cost of the sealer in 4 L containers = $6 \times \$55$

$$= \$330$$

Total cost of the sealer in 6 L containers = $4 \times \$87$

$$= \$348$$

It is cheaper to buy 6×4 L containers of sealer.

- What parts of this chapter do you remember from previous lessons?

- What careers use rates in their everyday work?

- How do you think you'll be able to use the concepts in this chapter in your life outside school?

- Are there any parts of this chapter that you don't understand? If yes, ask your teacher for help.

Copy and complete this mind map of the topic, adding detail to its branches and using pictures, symbols and colour where needed. Ask your teacher to check your work.

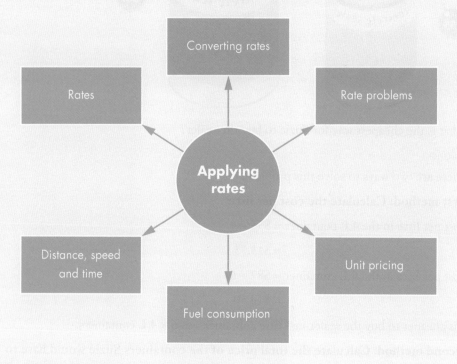

1 A tradesman charged $160 for $2\frac{1}{2}$ hours work. Calculate this rate in dollars/hour.

Exercise 7.01

2 David is driving at an average speed of 80 km/h. How far will he drive in:

   a 3 hours? **b** 15 minutes?

Exercise 7.02

3 The Japanese bullet train travelled 40 km in $7\frac{1}{2}$ minutes. Calculate the train's average speed in km/h.

Exercise 7.02

4 Convert a speed of 75 km/h into m/s. Answer to one decimal place.

Exercise 7.03

5 The Earth travels at approximately 30 km/s in its orbit around the Sun. What is this speed in km/h?

Exercise 7.03

6 The same brand of olive oil is available in 3 sizes.

Exercise 7.04

	Size	Cost
Small	375 mL	$5
Medium	750 mL	$8
Large	1 L	$13

   a Calculate the cost per 100 mL in each size.

   b Which size is the best value for money?

7 Denis charges $55/m² to lay marble tiles. How much will he charge for laying 42 m² of marble tiles?

Exercise 7.05

8 A tradesman charges a $50 callout fee which includes the first 15 minutes. For calls longer than 15 minutes, he charges an additional $28 per 15 minutes or part thereof. How much will the tradesman charge for a callout that takes 45 minutes?

Exercise 7.05

9 Olivia has been offered a new job. She has to choose her pay rate:
$36.50 per hour or a flat rate of $1400 per week.
In the job, Olivia will be working 40 hours per week. Which pay rate is the better deal?

Exercise 7.05

10 Kim is comparing the fuel economy of 2 cars.

Exercise 7.06

Car	Distance travelled	Fuel used
White car	620 km	53 L
Red car	470 km	43 L

   a Calculate the rate at which both cars used fuel in L/100 km. Answer to one decimal place.

   b Which car is the more fuel-efficient?

MEASUREMENT

8.

FINDING THE RIGHT PATH

Chapter problem

Isiah is the project manager for a construction company. His company is planning to construct 7 holiday cabins in a wilderness area. Isiah's task is to decide where to position the underground electricity cables and gas pipes using the smallest quantity of cables and pipes.

The diagram on the next page shows the positions of the 7 cabins and the distances between them. What is the minimum quantity of cables and pipes Isiah will need to connect each cabin to the main supply that is located 100 m from cabin 1? How should he position the cables and pipes?

CHAPTER OUTLINE

Cabin 2

Cabin 1

Cabin 3

60 m

55 m

Main supply 100 m

25 m

70 m

40 m

Cabin 5

50 m

30 m

20 m

35 m

Cabin 4

40 m

25 m

Cabin 6

Cabin 7

WHAT WILL WE DO IN THIS CHAPTER?

- Learn about network diagrams and terminology
- Solve problems involving networks, such as planning trips and activities
- Find the shortest path or fastest time between locations using Prim's or Kruskal's algorithms for minimum spanning trees

HOW ARE WE EVER GOING TO USE THIS?

- When we're planning a journey or activity
- When we're trying to minimise time or supplies to complete a task
- Project managers use networks to plan and organise projects
- Computer, gas and electricity installers require techniques to minimise cables and pipes

8.01 Networks

A **network** diagram shows connections or links between objects, such as the electrical wiring in a home, the roads joining a group of towns, a rail network, and even friendships on a social network such as Facebook.

This is a road network linking 4 towns, A, B, C and D, with the lengths of the roads shown in kilometres.

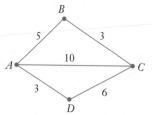

In a network, the lines are called **edges** and the points are called **vertices** (plural of **vertex**). When there are numbers on the edges, they are called **weighted edges**. Weighted edges can represent distance, time or money.

A **path** in a network is a route where there are no repeated edges or vertices. For example, in the road network above, $BCDA$ is a path but $ABCAD$ is not a path, because A appears twice. A **closed path** begins and ends at the same vertex, for example, $DCAD$. An **open path** begins and ends at different vertices, such as $CDAB$.

A network does not have to be drawn to scale. In the road network above, the 3 km edge BC looks longer than the 6 km edge DC. Also, real-life roads can be curved and go up and down hills, but in a network diagram we draw them as straight lines. The same road network could also be drawn like this:

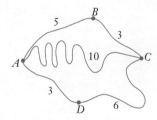

We can use network diagrams to solve problems such as finding the shortest route between two places, or determining whether it's possible to visit all places in the network by covering each road just once. We call this 'traversing a network'.

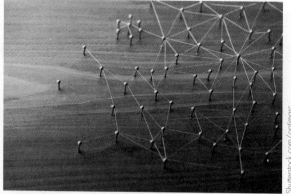

Shutterstock.com/optimarc

The degree of a vertex

The **degree** of a **vertex** is the number of edges that meet at the vertex.

In the road network diagram on the previous page, vertex A has degree 3 and vertex B has degree 2.

An **even vertex** has an even degree, and an **odd vertex** has an odd degree.

> If you can **traverse** a network, it means that you can visit all **vertices** by travelling along every edge exactly once.
>
> **Traversing** a network is only possible if all its vertices are even, or it has exactly two odd vertices (all the other vertices are even). We start at one odd vertex and finish at the other.

EXAMPLE 1

a For this road network, list the 3 paths to travel from A to C and the length of each path. The distances shown are in kilometres.

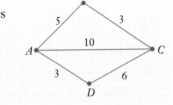

b Which path from A to C is the shortest?

c What is the degree of each vertex?

d Why is this network traversable?

e Which route could we travel to visit all 4 towns and drive on each road once only?

Solution

a List the 3 paths and their distances.

Path AC (directly): 10 km
Path ABC: $5 + 3 = 8$ km
Path ADC: $3 + 6 = 9$ km

b Find the shortest path.

The shortest path from A to C is ABC, via B.

c The degree of a vertex is the number of edges that meet at the vertex. Simply count all the edges.

Degree of vertex $A = 3$
Degree of vertex $B = 2$
Degree of vertex $C = 3$
Degree of vertex $D = 2$

d

The network is traversable because it has exactly two odd vertices, A and C.

e The path begins and ends on an odd vertex, so it must be A and C.

One possible route is $ABCADC$.

Exercise 8.01 Networks

In this exercise, all measurements on network diagrams are in kilometres.

1 For each network, count the number of vertices and the number of edges.

a

b

c

d

e

f

g

h

i

Example 1

2 a List the 3 paths from *A* to *D*.
b Calculate the length of each path from *A* to *D*.
c Which path from *A* to *D* is the shortest?
d List the degree of each vertex in the network.
e Can we traverse this network?

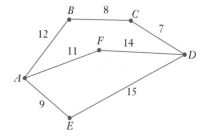

3 a List 3 possible ways to travel from *A* to *B* and the distance involved with each route.
b How long is the path from *A* to *B* via *C* and *D*?
c What is the shortest distance from *A* to *B*?
d Which vertices have a degree of 3?

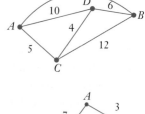

4 a List 4 ways you could travel from *A* to *C*.
b Which one of the 4 ways is the longest?
c Describe how you could traverse this network.

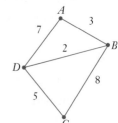

ISBN 9780170413534

5 a List all the routes from A to C and determine which of the routes is the shortest.

b Which vertices have an odd degree?

c Can you traverse this network?

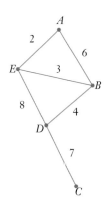

6 a What is the shortest route from B to D in this network?

b Which vertex has a degree of 4?

c Explain why it is impossible to visit every vertex in this network without travelling on some roads more than once.

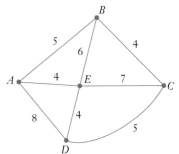

7 Jack is visiting Margaret River in WA. He sketched a network drawing of the places he wants to visit.

a Explain why it is possible for Jack to visit each of the locations and travel each road exactly once.

b Where can Jack start his tour?

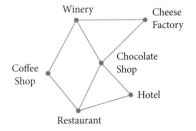

8 To help fire trucks reach fires efficiently, the local rural fire station has a network diagram of the area.

There are fires at The Ridge and One Hill Lookout. Describe the shortest routes from the fire station for the fire trucks to reach the fires.

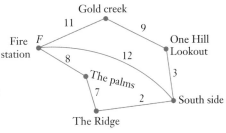

INVESTIGATION

MY NETWORK DIAGRAM

Construct a network diagram showing paths from your classroom to the school canteen. Include the distances in metres. Identify the shortest path from the classroom to the canteen.

8.02 The shortest path

Listing all the paths in a network diagram and calculating their lengths isn't a very efficient way to find the **shortest path** between two points. Fortunately, there is a logical process we can follow to eliminate a lot of the calculations.

EXAMPLE 2

Find the length of the shortest path from *A* to *C* in this network. The distances shown are in kilometres.

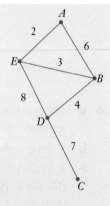

Solution

Starting at *A* we can choose to travel to *E* or *B*. Write the distance next to each vertex.

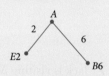

We could also travel to *B* via *E*. It is a shorter 5 km path from *A* to *B* via *E*. Cross out the 6 and replace it with 5.

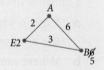

Now consider the path from *A* to *D* via *E* or *B*. From *E* it is 2 + 8 = 10, but from *B* it is only 5 + 4 = 9 to *D*. The shortest path from *A* to *D* is 9.

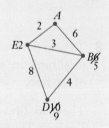

The last step is the section from *D* to *C*. There is only one path: 9 + 7 = 16.

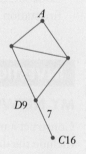

Write your answer.

The shortest path from *A* to *C* is 16 km. The path is *AEBDC*.

ISBN 9780170413534

Exercise 8.02 The shortest path

In this exercise, all measurements on network diagrams are in kilometres.

1 Follow these steps to determine the shortest path from *A* to *C*.

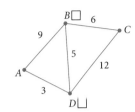

a Copy this diagram and write the distances from *A* to *B* and *A* to *D* in the appropriate square next to *B* and *D*.

b Now think about getting to *B* via *D*. Explain why we cross out the 9 and replace it with an 8.

c Use the 8 at *B* and the 3 at *D* to calculate the shortest distance to *C*.

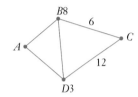

d Describe the shortest path from *A* to *C*.

2 Follow these steps to determine the shortest path from *A* to *C*.

a Copy this diagram and place appropriate values in the squares at *B* and *D*.

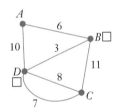

b Think about the values in the squares at *B* and *D*.
Is there a shorter route via the path between *B* and *D*? If there is, cross out the old value and write in the shorter one.

c It's shorter to go from *D* to *C* via the curved route. Add 7 to the value in the square next to *D*, then compare it to the value you get when you add 11 to the value in the square next to *B*. Write the smaller number in the square next to *C*.

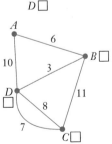

d What is the shortest path from *A* to *C*? How long is it?

3 Determine the shortest path from *A* to *F* in this network diagram.

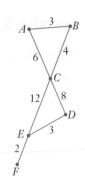

4 Tran is walking through this network from *A* to *D*.

   a What is the shortest path he can follow?

   b Tran walks at 4 km/h. He is leaving point *A* at 9:30 a.m. What is the earliest time he will arrive at *D*?

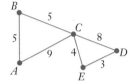

5 Toulla is planning a bushwalk starting at *A* and finishing at *D*.

   a Describe the shortest route from *A* to *D*.

   b At an average speed of 4.5 km/h, how long will it take Toulla to walk the shortest route from *A* to *D*?

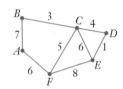

6 The shortest route from *A* to *C* in this network is 30 km long. How long is the section from *E* to *B*?

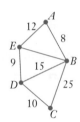

7 In this network, what could be the lengths of sections *AB*, *AD* and *BC* if the shortest route from *A* to *C* is *ABDC*?

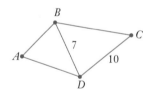

DISTANCE, TIME OR MONEY?

Every day, Kate travels to work from home and back again. The network diagrams show the paths that she can take. The first shows the distances, the second shows the times and, as some of the roads involve tolls, the third shows the costs involved.

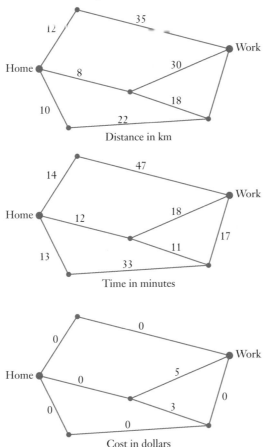

Distance in km

Time in minutes

Cost in dollars

Your task is to decide which route Kate should take to and from work. Should she take the same route home as she takes to work? What route do you think her GPS would give her? Kate starts work at 8:45 a.m. At what time should she leave home in the morning? Calculate her weekly toll cost if you decide to use toll roads.

8.03 Maps and networks

EXAMPLE 3

At the school camp, students are accommodated in 3 cabins, labelled *A*, *B* and *C*. The camp site also contains a meal hall, a shower block and a sports complex. Some of the buildings are connected by paths as shown in the table below. All measurements are in metres. Calculate the distance along covered paths from Cabin *C* to the Sports complex.

	Cabin *A*				
Cabin *B*	–	Cabin *B*			
Cabin *C*	–	25	Cabin *C*		
Meal hall	15	10	–	Meal hall	
Shower block	–	–	–	12	Shower block
Sports complex	–	–	–	20	40

Solution

Use the table to draw a network diagram. Cells marked – in the table mean that the pair of buildings is *not* connected by a path. The 15 in the cell connecting the meal hall with Cabin *A* means that there is a 15 m path between the two buildings. Join these two buildings and show the 15 m on the join.

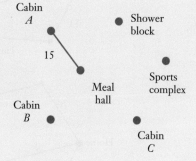

Use the information in the table to connect all the other buildings that have covered paths joining them.

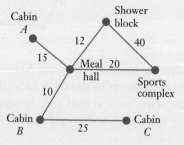

Students who want to travel from Cabin *C* to the sports complex along paths have to go via Cabin *B* and the meal hall.

Distance $= 25 + 10 + 20$
$= 55$ m

ISBN 9780170413534

Exercise 8.03 Maps and networks

1 Binnsville Botanical Gardens contains 5 gardens connected by paths as shown in the table. All measurements are in metres.

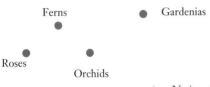

	Orchids			
Ferns	70	Ferns		
Gardenias	80	50	Gardenias	
Native plants	–	–	40	Native plants
Roses	–	35	–	90

a Use the table to copy and complete this network diagram for the botanical gardens.

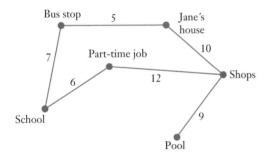

b How far is it along paths from the Gardenias to the Roses?

c Describe 3 routes you could take to walk along paths from the Native plants to the Orchids.

d How long is the shortest route along paths from the Native plants to the Orchids?

2 a Copy and complete the table for this network diagram. The numbers represent walking time in minutes.

	Bus stop	Jane's house	Part-time job	Pool	School
Jane's house					
Part-time job					
Pool					
School					
Shops					

b How long does it take to walk to school from Jane's house along the fastest route?

c Suggest a reason why Jane might choose not to use the fastest route to walk to school.

3 This diagram shows the Singapore train network.

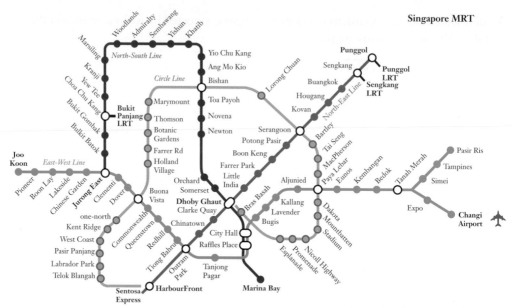

Based on: Land Transport Authority Singapore MRT

a Hubert caught a train at Harbour Front on the blue line. He changed to the green line at Outram Park, then changed to the red line at Jurong East. He got off the train 3 stops later. Where did he get off the train?

b Jayanthi is staying near Orchard station on the red line and she needs to travel to Changi Airport to catch her flight home. Describe an efficient route she could follow to travel from Orchard Station to Changi Airport.

ISBN 9780170413534

4 Peta is visiting a national park and she plans to hike from the car park to the beach, taking the shortest path. Use the information in the table below to determine:

a the length of the hike

b the approximate time the hike will take her at an average speed of 3.5 km/h.

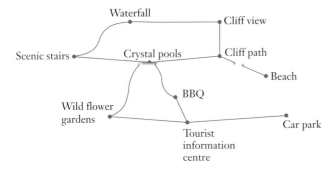

Distances along tracks in the national park are in metres.

	BBQ	Beach	Car park	Cliff path	Crystal pools	Scenic stairs	Tourist Information centre
Cliff path	–	840	–				
Crystal pools	1700	–	–	900			
Scenic stairs	–	–	–	–			
Tourist information centre	60	–	200	–	–	–	
Waterfall	–	–	–	–	–	865	–
Wild flower gardens	–	–	–	–	740	–	485

INVESTIGATION

SINGAPORE TRIP

Imagine you are about to visit Singapore for the first time. You are going to fly into Changi Airport, stay at a hotel in Newton and then travel around the city by train. During the trip you want to visit the following places:

- Outram Park
- Botanic Gardens
- Little India
- City Hall
- Queenstown
- Harbour Front

Use the train network diagram on page 232 to plan your route.

8.04 Directed networks

All the networks we've used so far are undirected networks because we can travel in either direction along the edges. **Directed networks** include 'one-way paths' that can be travelled in one direction only.

EXAMPLE 4

This directed network shows one food web in the Australian bush. The arrows point from the food item to the animal that eats it.

According to this food web:

a what do kookaburras eat?

b what eats kookaburras?

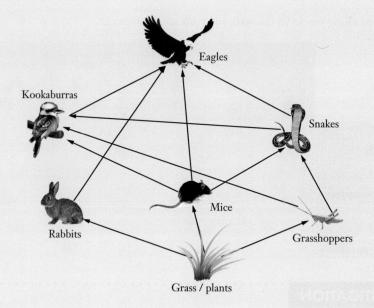

Solution

a The arrows pointing towards the kookaburra indicate its food source.

Kookaburras eat mice, snakes and grasshoppers.

b The arrow pointing away from the kookaburra shows what eats kookaburras.

Eagles eat kookaburras.

ISBN 9780170413534

EXAMPLE 5

This directed network shows one approach to completing a research assessment task.

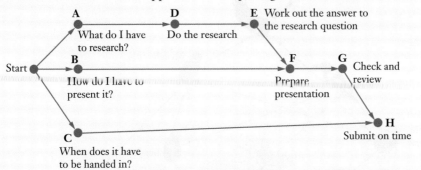

Use the network diagram to complete the activity table.

Activity	Prerequisite
Start	
A	
B	
C	
D	
E	
F	
G	
H	

A prerequisite to an activity is another activity that must be completed before the activity can start.

Solution

Don't write anything next to 'Start': no prerequisite.
Activities A, B and C are starting activities, so write 'Start' next to them.

We have to complete A before we can do D.
For D, the prerequisite is A.
We have to complete D before E, so for E, the prerequisite is D.
We have to complete both E and B before F.
F has 2 prerequisites, E and B.

Complete the remainder of the table in a similar way.

Activity	Prerequisite
Start	
A	Start
B	Start
C	Start
D	A
E	D
F	E, B
G	F
H	G, C

Exercise 8.04 Directed networks

Example 4

1 This directed network diagram shows a food web in an Australian garden.

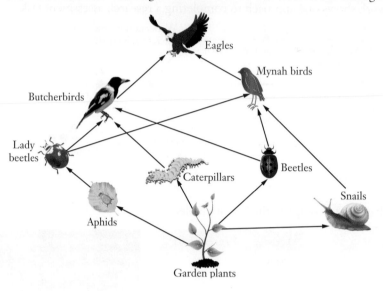

According to the diagram:

a what eats aphids?

b what do butcherbirds eat?

c What would be the likely effect if there were no lady beetles in the garden?

Example 5

2 Match each directed network in parts **a** to **d** with its activity table in parts **i** to **iv**.

a

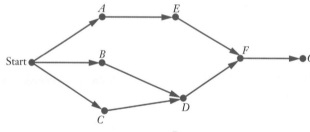

b

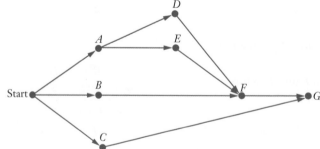

c

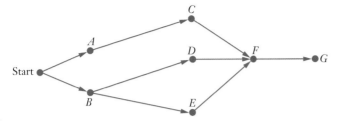

d

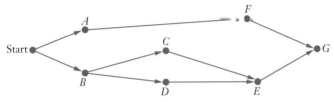

i

Activity	Prerequisite
Start	
A	Start
B	Start
C	B
D	B
E	C, D
F	A
G	E, F

ii

Activity	Prerequisite
Start	
A	Start
B	Start
C	start
D	A
E	A
F	B, D, E
G	C, F

iii

Activity	Prerequisite
Start	
A	Start
B	Start
C	Start
D	B, C
E	A
F	D, E
G	F

iv

Activity	Prerequisite
Start	
A	Start
B	Start
C	A
D	B
E	B
F	C, D, E
G	F

3 Construct a directed network for each project.

a

Activity	Prerequisite
A	Start
B	A
C	A
D	A
E	B
F	C
G	E, F, D

b

Activity	Prerequisite
A	Start
B	A
C	B
D	B
E	C
F	D
G	E, F

c

Activity	Prerequisite
A	Start
B	A
C	B
D	C
E	D
F	D
G	E, F

d

Activity	Prerequisite
A	Start
B	A
C	A
D	B
E	B
F	C
G	D, E, F

4 Recently, Brad and David paid off their block of land on which they are going to build a house. Their bank has agreed to lend the money they need for the house but they must first determine their costs. This is an alphabetical list of the steps they need to complete.

- Apply for the loan (*A*)

- Build the house (*B*)

- Choose a builder (*C*)

- Choose a plan (*D*)

- Get quotes from 3 builders (*E*)

- Move in (*F*)

- Obtain council approval for the building (*G*)

- Select items such as doors, windows, tiles, taps and floor coverings (*H*)

Construct a directed network showing the steps Brad and David need to follow.

5 Group challenge! The directed network and the table show the majority of tasks involved in organising the school musical. Copy the network diagram and match the labels *A* to *L* in the network diagram with the tasks in **i** to **xii** in the table. Task **i** is already in position.

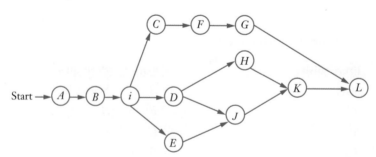

i	Copyright approval to perform the show	vii	Orchestra auditions
ii	Choose the musical (show)	viii	Practice (rehearse)
iii	Design tickets	ix	Print tickets
iv	Dress rehearsal	x	Select the organising committee
v	Make costumes	xi	Sell tickets
vi	Opening night	xii	Singer auditions

INVESTIGATION

MOVING OUT

Imagine that you are going to move out of home and rent a place of your own.
What steps and decisions would you have to make before you can do it?
Construct a directed network diagram to show your action plan.

8.05 Spanning trees

A **tree** is a network in which any two vertices are connected by exactly one path. It does not contain any cycles where you can return to the same vertex along a different edge.

The following networks are trees.

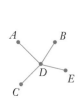
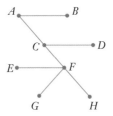

The following networks are not trees.

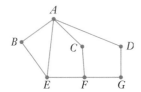

For any network, a **spanning tree** is a tree that connects every vertex of the network. There are lots of different spanning trees for every network.

EXAMPLE 6

Construct a spanning tree for this network.

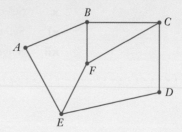

Solution

Choose any vertex, say *D*. It doesn't matter which one we choose!
Then join *D* to either *C* or *E*. Let's choose *C*.

Now we can join *C* to *F* or *B*
OR join *D* to *E*.
Let's choose *C* to *F*.

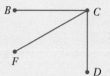

Now we can join *F*, *C* or *D* to any unused vertex.
Let's choose *C* to *B*.

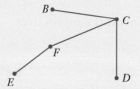

Now we can join any of the vertices we've used to any unused vertex, provided we don't make a cycle. Let's join *F* to *E*.

Now we can join any of the vertices we've used to any unused vertex, provided we don't make a cycle. Let's join *F* to *E*.

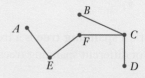

A is the only vertex left to join. We can join either *B* or *E* to *A*. Let's join *E* to *A*.
This is a spanning tree: all vertices connected without making a cycle.

The number of **edges** in a **spanning tree** is one less than the number of **vertices** in the original network:

Number of edges in spanning tree = Number of vertices in original network − 1

Checking for Example 6: Number of vertices in original network = 6, so we should have 6 – 1 = 5 edges in the spanning tree. ✓

These diagrams show two other spanning trees for the network in Example 6.

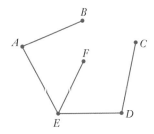
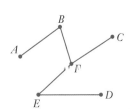

EXAMPLE 7

Explain why each of the following diagrams can't be spanning trees for the network in Example 6.

a
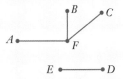

b
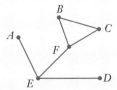

c
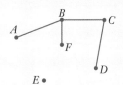

d
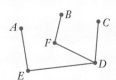

Solution

a   There are two separate sections. Joining *E* to *F* would fix the problem.

b   There is a cycle, *FBC*. Removing any one of *BC*, *BF* or *CF* would fix the problem.

c   Vertex *E* isn't connected. *E* could be connected to any one of vertices *A*, *F* or *D* to fix the problem.

d   The edge *FD* is illegal as it isn't an edge in the original network. Remove *FD* and join either *AB* or *BC* to fix the problem.

Exercise 8.05 Spanning trees

1 Explain why each diagram can't be a spanning tree.

a   **b**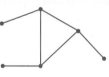

c   **d**

2 a Construct two different spanning trees for this network.

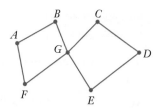

b Explain why this diagram cannot be a spanning tree for the network in part **a**.

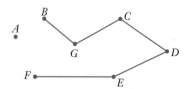

3 Damien drew two different spanning trees for one network diagram.

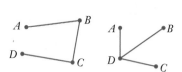

a Draw what the original network could look like.

b Draw two more possible spanning trees for the network.

4 This network with 3 vertices and 3 edges can have three possible different spanning trees. Draw all three possible spanning trees.

5 A network with n vertices has n^{n-2} possible spanning trees.

a Use the formula to show that there are 16 possible spanning trees for this network diagram.

b How many of the 16 possible spanning trees can you draw?

8.06 Prim's algorithm for minimum spanning trees

Minimum spanning trees

A **minimum spanning tree** for a network is the spanning tree whose edges have the lowest weights. For a road network, it is the spanning tree that gives the shortest total distance when you add up the edges. To find the minimum spanning tree for a network, we can use a repetitive method called **Prim's algorithm**, which starts at any vertex and then chooses the shortest edge to link to another vertex.

EXAMPLE 8

Prim's algorithm

Use Prim's algorithm to construct the minimum spanning tree for this network.

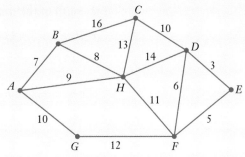

Solution

Choose any vertex as a starting point.

Choose C as the starting point.

Select the shortest edge from C to another vertex. (10)

Select the shortest edge from C and D to another vertex. (3)

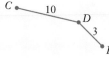

Select the shortest edge from C, D and E to another vertex. (5)

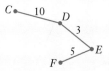

Select the shortest edge from C, D, E and F to a vertex not already used. It can't be 6 because we've used D already and joining F to D would make a cycle, which isn't allowed. (11)

Choose the shortest edge from C, D, E, F and H to a new vertex. (8)

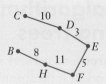

The next shortest edge to a new vertex is from B to A. (7) Similarly, the shortest edge to G is from A. (10)

The minimum spanning tree is complete.

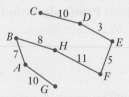

Length of the minimum spanning tree

$= 10 + 3 + 5 + 11 + 8 + 7 + 10$

$= 54$

Check! There are 8 vertices.

Edges = $8 - 1 = 7$ ✓

There are 7 edges.

Exercise 8.06 Prim's algorithm for minimum spanning trees

Example 8

1 Use Prim's algorithm to construct a minimum spanning tree for each network.

a

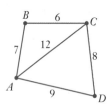

b

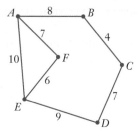

c

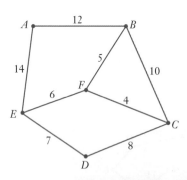

d

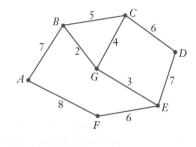

> When you're constructing a minimum spanning tree it's a good idea to cross out the vertices as you use them. Then you won't accidentally use the same vertex twice.

2 a Construct a minimum spanning tree (MST) for the network in Example 8, starting at vertex H.

b Is your MST the same as the MST in the example solution?

3 A country hospital contains 6 separate buildings labelled A to F connected by walkways. For night-time security, management is going to put floodlights on some of the paths. Management wants the cost to be minimised, but insists that it must be possible to walk to every building along a lighted route. The diagram shows the cost, in dollars, of lighting each path.

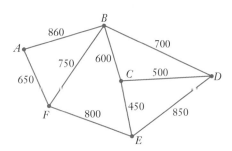

a Construct a minimum spanning tree for this network.

b Calculate the minimum cost for suitable lighting.

4 A small university campus is going to install high-speed Internet cables to its 9 buildings, labelled A to I. The network diagram shows the distances in metres between the buildings.

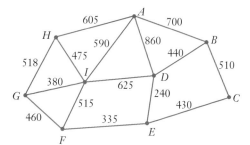

a Construct a minimum spanning tree for the network.

b What is the minimum length of cable required?

5 Solomon needs to supply electricity to some components labelled P to V in an electrical circuit he is building. The network diagram shows the lengths of the wires in millimetres.

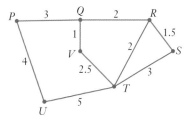

Calculate the minimum amount of wire he will need.

8.07 Kruskal's algorithm for minimum spanning trees

Another repetitive method for finding minimum spanning trees is **Kruskal's algorithm**, which starts with the shortest edge, then the next shortest edge, and so on. It works differently to Prim's algorithm, but it produces the same minimum spanning tree.

EXAMPLE 9

Use Kruskal's algorithm to construct the minimum spanning tree for this network. It is the same network diagram as in Example 8 on page 243.

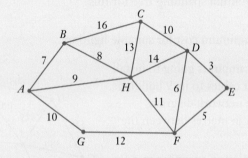

Solution

Start with the smallest edge. (3)

D • —3
• E

Find the next smallest edge. (5)

Find the next smallest edge to a new vertex.
It can't be *FD* (6) because we have used vertex *D* already and that would make a cycle. (7)

B •   D • —3
7 /   • E
A •   5 /
            • F

Find the next smallest edge. (8)

B •   D • —3
7 / 8   • E
A • • H   5 /
            • F

The next smallest edges are 10 for *CD* and *AG*. It can't be 9 for *AH* because we've used *H* already.

C • —10
B •   • D
7 / 8   3 •
A • • H   • E
10 \   5 /
 G •   • F

The next size is 11.

The minimum spanning tree is complete, and it's the same minimum spanning tree as in Example 8 using Prim's algorithm.

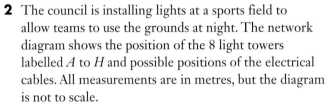

Check! There are 8 vertices.   There are 7 edges.

Edges = 8 − 1 = 7 ✓

Exercise 8.07 Kruskal's algorithm for minimum spanning trees

1 Use Kruskal's algorithm to construct the minimum spanning trees for each network in Exercise 8.06 Question **1** on page 244.

Example
9

2 The council is installing lights at a sports field to allow teams to use the grounds at night. The network diagram shows the position of the 8 light towers labelled *A* to *H* and possible positions of the electrical cables. All measurements are in metres, but the diagram is not to scale.

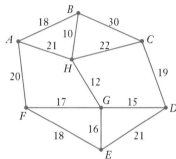

  a Use Kruskal's algorithm to construct a minimum spanning tree for the electrical cables for the lights.

  b How many metres of electrical cable will be required for the lights?

3 Tom wants to supply underground irrigation to the 10 trees in his garden. The network diagram shows the possible positions of irrigation pipes between the trees and the distances in metres. He labelled the trees *A* to *I*.

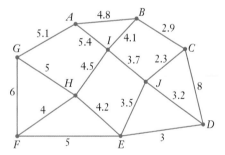

  a Use Kruskal's algorithm to construct a minimum spanning tree.

  b Calculate the minimum length of irrigation pipes that Tom will need.

4 The paths in a large orchard often become unpassable in wet weather. The owner wants to upgrade the paths to allow for all-weather access. The network diagram shows the current paths between each fruit-holding area (labelled *A* to *J*) and the table shows the distances along each path.

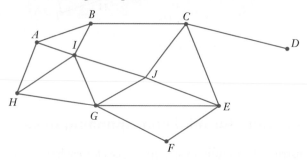

	A	B	C	D	E	F	G	H	I	J
A		110	–	–	–	–	–	75	65	–
B	110		82	–	–	–	–	–	89	–
C	–	82		94	135	–	–	–	103	98
D	–	–	94		–	–	–	–	–	–
E	–	–	135	–		36	116	–	–	56
F	–	–	–	–	36		72	–	–	–
G	–	–	–	–	116	72		47	61	53
H	75	–	–	–	–	–	47		64	–
I	65	89	103	–	–	–	61	64		38
J	–	–	98	–	56	–	53	–	38	

a The plan is to upgrade the paths to allow access along upgraded paths to each fruit-holding area with the minimum path length. Copy and complete the network diagram above, then use Kruskal's algorithm to calculate the minimum path length.

ISBN 9780170413534

b The construction team quoted different rates for each path. The quote, in dollars, for each section is shown on the following network diagram. Calculate the minimum cost to upgrade the paths.

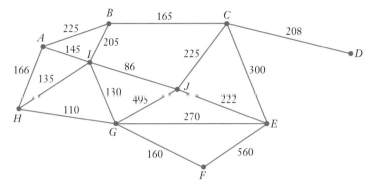

c Suggest a reason why upgrading some of the paths is more expensive than their length would indicate.

KEYWORD ACTIVITY

SIMILAR WORDS

In this chapter, there are several pairs of words that have a similar but slightly different meaning. Explain, in the context of this chapter, what each word in the pair has in common and how they are different.

1 Vertex – vertices

2 Edges – weighted edges

3 Directed network – undirected network

4 Spanning tree – minimum spanning tree

5 Prim's algorithm – Kruskal's algorithm

Networks
find-a-word

SOLUTION TO THE CHAPTER PROBLEM

Problem

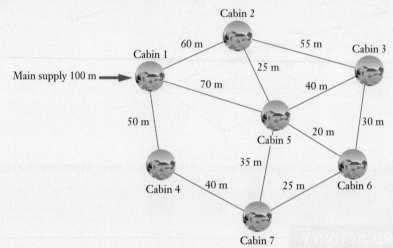

Isiah is the project manager for a construction company. His company is planning to construct 7 holiday cabins in a wilderness area. Isiah's task is to decide where to position the underground electricity cables and gas pipes using the smallest quantity of cables and pipes.

The diagram above shows the positions of the 7 cabins and the distances between them. What is the minimum quantity of cables and pipes Isiah will need to connect each cabin to the main supply that is located 100 m from cabin 1? How should he position the cables and pipes?

Solution

Use Prim's or Kruskal's algorithm to construct the minimum spanning tree.

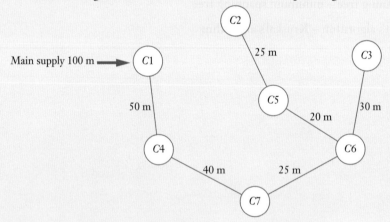

Total amount of cable and pipes = 100 + 50 + 40 + 25 + 30 + 20 + 25

$$= 290 \text{ m}$$

Isiah will require 290 m each of electricity cable and gas pipes.

- Where have you seen a network diagram in your life outside school?
- What application of networks do you think is the most interesting?
- Who might use a directed network in their job?
- Is there any part of this chapter that you don't understand? If yes, ask your teacher to explain it to you.

Copy and complete this mind map of the topic, adding detail to its branches and using pictures, symbols and colour where needed. Ask your teacher to check your work.

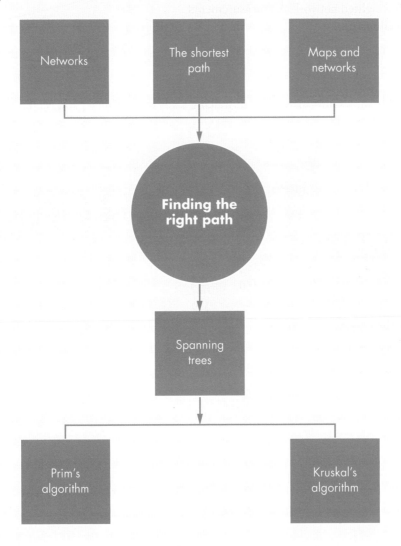

1 a How many vertices are there in this network?

b How many edges are there?

c What is the degree of vertex D?

d What is the degree of vertex E?

e Which vertices in the network are odd?

f Is it possible to traverse this network? If yes, where do you start and finish?

2 Calculate the shortest route from A to F in this weighted network. All measurements are in kilometres.

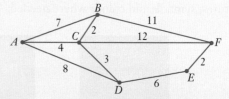

3 The table shows information about walking trails in a national park. All distances are in metres.

	Ancient rock paintings	Cliff view	Entrance	Fossils	Wollemi Pine grove	Waratah gardens	Waterfall
Ancient rock paintings		–	–	180	275	–	–
Cliff view	–		160	–	–	–	140
Entrance	–	160		350	–	–	–
Fossils	180	–	350		–	–	–
Wollemi Pine grove	275	–	–	–		–	–
Waratah gardens	–	–	–	–	–		200
Waterfall	–	140	–	–	–	200	

a Copy and complete the network diagram to show the walking trails in the park.

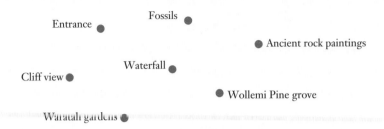

b David wants to photograph the Wollemi Pines. How far will he have to walk from the entrance and what route should he follow?

Alamy Stock Photo/Wildlight Photo Agency

ISBN 9780170413534

4 This directed network shows the stages for completing a design project.

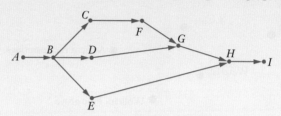

a What is the last task that has to be completed before the project is finished?

b Which task must be completed before task E can start?

c Copy and complete this activity table for the design project.

Task	Prerequisite
A	Start
B	
C	
D	
E	
F	
G	
H	
I	

5 Why aren't the following diagrams spanning trees for this network?

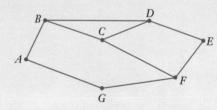

a

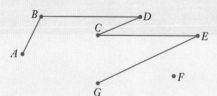

b

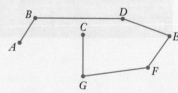

c

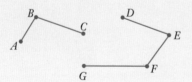

d

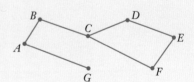

Exercise
8.05

Exercise
8.06

Exercise
8.07

6 Construct two different spanning trees for the network in Question **5**.

7 Use Prim's algorithm to construct the minimum spanning tree for this network.

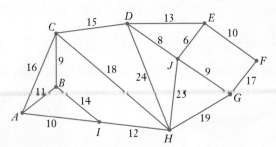

8 The managers of a holiday campsite want to install a sewerage system connected to each cabin. The cabins are labelled *A* to *I*. Use Kruskal's algorithm to determine the minimum spanning tree connecting the cabins and the minimum length of pipe required for the sewerage system.

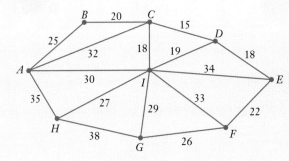

9.

SO YOU'VE GOT A RIGHT ANGLE

Chapter problem

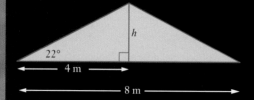

Denis is building a garage. The garage is 8 m wide and the pitch of the roof is 22°. How high does he need to make the truss?

CHAPTER OUTLINE

Note: Bearings and angles of elevation and depression are covered in Chapter 11.

WHAT WILL WE DO IN THIS CHAPTER?

- Solving problems using Pythagoras' theorem
- Use trigonometry to calculate the lengths of sides in right-angled triangles, including the hypotenuse
- Use trigonometry to calculate the sizes of angles in right-angled triangles, in degrees and minutes

HOW ARE WE EVER GOING TO USE THIS?

- Determining a length when we can't measure it
- Many people who work in trades, for example, carpenters and builders, use right-angled triangle calculations in their work

Pythagoras'
theorem

Pythagoras'
leopard

Pythagoras'
problems

Pythagoras'
theorem time trial

Pythagorean
two-step problems

Applications of
Pythagoras'
theorem

9.01 Pythagoras' theorem

Pythagoras was an ancient mathematician who lived in Greece from 580 to 500 BCE. He was the founder of a secret brotherhood who was interested in the mystical properties of numbers. **Pythagoras' theorem** is named after him even though no-one is sure whether it was Pythagoras himself or another member of the brotherhood who proved the theorem. What we do know is that ancient scholars from different civilizations knew about the Pythagorean formula well before Pythagoras himself.

Pythagoras' theorem

In a right-angled triangle, the square of the hypotenuse is equal to the sum of the squares of the other two sides.

$(\text{hypotenuse})^2 = (\text{side 1})^2 + (\text{side 2})^2$

$c^2 = a^2 + b^2$

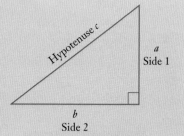

The **hypotenuse** is the longest side of a right-angled triangle.

EXAMPLE 1

Use Pythagoras' theorem to find the height (h) of this roof truss, correct to one decimal place.

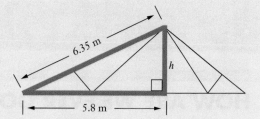

Solution

The hypotenuse is 6.35 and one side is 5.8.

Write Pythagoras' theorem.

Solve the equation for h.

$$c^2 = a^2 + b^2$$
$$6.35^2 = 5.8^2 + h^2$$
$$h^2 = 6.35^2 - 5.8^2$$
$$= 6.6825$$
$$h = \sqrt{6.6825}$$
$$= 2.5850 \ldots$$
$$\approx 2.6 \text{ m}$$

We can also use Pythagoras' theorem to test whether a triangle contains a right angle. If the theorem works, there's a right angle. If the theorem doesn't work, the triangle isn't right-angled.

EXAMPLE 2

Is this triangle right-angled?

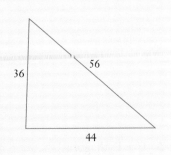

Solution

We need to check whether $56^2 = 36^2 + 44^2$.

If both sides of the equation are equal, then there is a right angle. If they aren't equal, there is no right angle.

$56^2 = 36^2 + 44^2$

$3136 = 1296 + 1936$

$3136 = 3232$

This statement is false.

Pythagoras' theorem doesn't work, so the triangle is *not* right-angled.

Exercise 9.01 Pythagoras' theorem

1 The diagram represents a wheelchair ramp. The 36 m long ramp covers a horizontal distance of 35 m. Use Pythagoras' theorem to calculate the rise, h m, of the ramp. Answer correct to one decimal place.

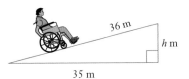

2 What length of wire is required to connect the top of a 23 metre TV antenna to a hook 6 metres from the base of the antenna? Answer correct to one decimal place.

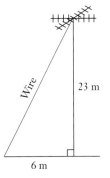

3 Gomez installs a new section of pipe to join 2 existing pipes. Calculate the length of the new section of pipe. Express your answer in metres correct to the nearest millimetre.

> When a measurement is in metres, the nearest mm means 3 decimal places. The nearest cm is 2 decimal places.

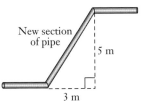

4 Bai is flying a small plane at 160 km/h against a 50 km/h wind, as shown in the diagram. Calculate the plane's ground speed correct to the nearest km/h.

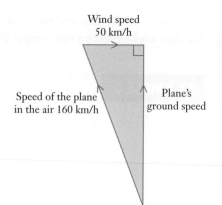

Wind speed 50 km/h

Speed of the plane in the air 160 km/h

Plane's ground speed

5 Use Pythagoras' theorem to test whether each triangle is right-angled.

a

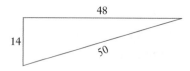

48

14

50

b

10

12.5

7.5

c

6

28

24

d

4.5

20.5

20

6 Mohammed is laying a concrete slab 12 m by 3 m in front of his shed. He uses Pythagoras' theorem to check that the corners of the slab are right angles. How long should the diagonal be? Express your answer in metres, correct to the nearest centimetre.

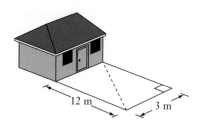

12 m

3 m

7 The school cross-country course is in the shape of a right-angled triangle. The first leg is 1200 m and the second leg is 950 m. The third leg, through thick scrub, is difficult to measure. Calculate its length, correct to the nearest metre.

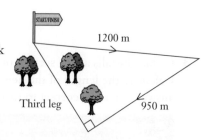

START/FINISH

1200 m

Third leg

950 m

8 Chan had to cross the river to get from Mud Springs to Eulo. The new road across the river was closed for repairs so he had to use the old bridge. How much further did he have to travel using the old road and old bridge compared to the direct route using the new road? Answer correct to one decimal place.

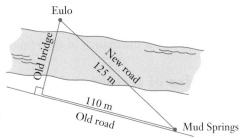

9 Joe is the pilot of a small plane. He planned to fly 200 km from Mount Surprise to Cairns. Because of poor weather conditions between Mount Surprise and Cairns, Joe flew 135 km due north to Chillagoe, then turned due east and flew to Cairns. Calculate the distance from Chillagoe to Cairns, correct to the nearest kilometre.

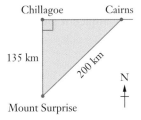

10 The hypotenuse of a right-angled triangle is 24 m. Draw 2 possible triangles, showing the lengths of the other 2 sides, correct to 2 decimal places.

INVESTIGATION

CUTTING RECTANGLES

Elspeth works in an upholstery business. One of her jobs is cutting out rectangular pieces of foam to make seat cushions. She always has a problem judging whether the cut foam is square (square means 'at right angles').

- Elspeth thinks the foam in the diagram is square. Is she right?
- Describe a process Elspeth could use to check whether her foam blocks are square.

9.02 The tangent ratio

Trigonometry is the branch of mathematics used to solve problems involving triangles.

The sides of a right-angled triangle

In a right-angled triangle, the longest side is the side opposite the right angle and it is called the **hypotenuse**. The names of the other two sides are determined by the reference angle.

The **opposite side** is the side facing the reference angle.

The **adjacent side** is next to the reference angle.

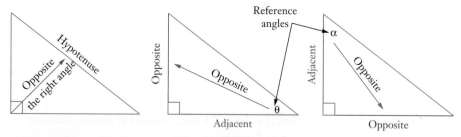

When naming the sides of a right-angled triangle, always start with the hypotenuse, because its position never moves. Then determine the opposite and adjacent sides.

The **tangent** ratio, abbreviated **tan**, is the ratio of the length of the **opposite side** to the length of the **adjacent side**.

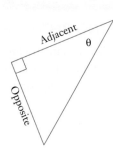

The tangent ratio

$$\tan\theta = \frac{\text{opposite}}{\text{adjacent}}$$

The tangent ratio can be used to calculate the length of a side or an angle in a right-angled triangle.

EXAMPLE 3

Use the tangent ratio to find the length of side x in the triangle. Express your answer correct to 2 decimal places.

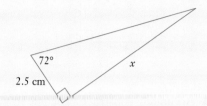

Solution

In the triangle, x is on the side opposite the angle 72° and the adjacent side is 2.5 cm long.

The tan ratio is $\dfrac{\text{opposite}}{\text{adjacent}}$. We have to make an equation, then solve it.

$$\tan 72° = \frac{x}{2.5}$$

$$\text{Or } \frac{x}{2.5} = \tan 72°$$

Multiply both sides by 2.5.

$$2.5 \times \frac{x}{2.5} = \tan 72° \times 2.5$$

$$x = 2.5 \tan 72°$$

Enter 2.5 [tan] 72 [=] on your calculator.

$$x = 7.69\,420 \ldots$$
$$\approx 7.69 \text{ cm}$$

From the diagram, $x \approx 7.69$ cm seems to be a reasonable answer.

> Make sure that your calculator is set to degrees mode DEG or D. If it is set to RAD or GRAD, your calculator will give you the wrong answer. Ask your teacher for help if you need.

Exercise 9.02 The tangent ratio

1 Copy each triangle and write the names of the sides with respect to angle θ.

a

b

c

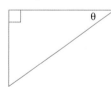

2 For this triangle, how long is:
   a the hypotenuse?
   b the side opposite α?
   c the side adjacent to β?

3 For each triangle, write tan θ as a fraction.

a
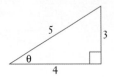

b
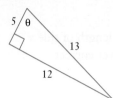

c
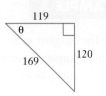

4 Accurately construct a triangle in which $\tan\theta = \dfrac{2}{3}$.

5 Copy and complete the working to calculate the value of t in the diagram. Express your answer correct to one decimal place.

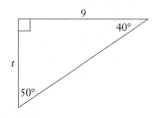

$$\frac{t}{9} = \tan\,\square$$

$$\square \times \frac{t}{9} = \tan\,\square \times \square$$

$$t \approx \square$$

6 Find the value of each pronumeral, correct to 2 decimal places.

a
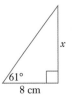

b
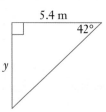

c
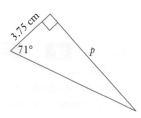

d
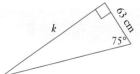

e

f
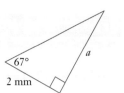

7 These 2 right-angled triangles share a side *BD*.
   a Use △*ABD* to find *h* correct to 2 decimal places.
   b Use △*BDC* to find *x* correct to one decimal place.

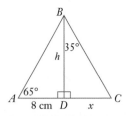

9.03 Using tan to find an angle

We measure angles in degrees, minutes and seconds in the same way as we measure time. There's 60 minutes (60') in a degree and 60 seconds (60") in a minute. If the seconds are 30 or greater, we round up the minutes.

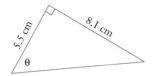

We can use the formula $\tan\theta = \dfrac{\text{opposite}}{\text{adjacent}}$ to calculate angles as well as sides.

EXAMPLE 4

Calculate the size of angle θ, correct to the nearest minute.

Solution

Use $\tan\theta = \dfrac{\text{opposite}}{\text{adjacent}}$ to find the angle θ.
The opposite side is 8.1 cm and the adjacent side is 5.5 cm.

$\tan\theta = \dfrac{8.1}{5.5}$

To 'undo the tan' we need to press
SHIFT tan on our calculator.

$\theta = 55.8230\ldots$

Enter SHIFT tan 8.1 ÷ 5.5 = .

Use the ○' " button on your calculator to express the answer in degrees and minutes.

$\theta = 55° \ 49' \ 22.84"$

Round down to the nearest minute because 22.84 is less than 30, or half a minute.

$\theta = 55° \ 49'$

---

1 Use your calculator to find the value of θ, correct to the nearest minute.

  a $\tan \theta = 0.86$ **b** $\tan \theta = 1.07$ **c** $\tan \theta = \dfrac{3}{4}$

  d $\tan \theta = \dfrac{3}{5}$ **e** $\tan \theta = \dfrac{11}{8}$ **f** $\tan \theta = 1\dfrac{1}{3}$

2 Use the tan ratio to find the value of each pronumeral correct to the nearest minute.

  a

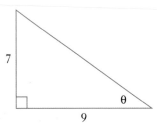

  b

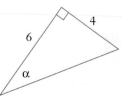

  c

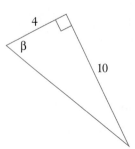

  d

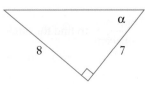

  e

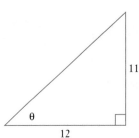

  f

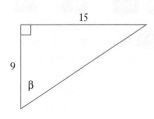

3 The shorter (non-hypotenuse) sides of a right-angled triangle are 4 cm and 6 cm long. Calculate the sizes of the angles in the triangle, correct to the nearest degree.

4 The shorter sides in a right-angled triangle are both 5 cm long.

  a   Find the sizes of the angles in the triangle.

  b   Explain why tan 45° = 1.

9.04 Sine and cosine

We use the **sine** and **cosine** ratios when calculations involve the hypotenuse. The abbreviation for sine is **sin** and it is pronounced 'sign'. The abbreviation for cosine is **cos**.

The sine and cosine ratios

$$\sin \theta = \frac{\text{opposite}}{\text{hypotenuse}}$$

$$\cos \theta = \frac{\text{adjacent}}{\text{hypotenuse}}$$

Trigonometric ratios

Trigonometric calculations

Identifying the correct trigonometric ratio

Finding an unknown angle

Finding an unknown side

Finding an unknown angle

EXAMPLE 5

What is the length of x in the triangle?
Express your answer correct to one decimal place.

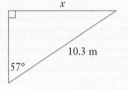

Solution

x is the opposite side, and 10.3 m is the hypotenuse. The opposite side and the hypotenuse are in the sin formula.

$$\sin 57° = \frac{x}{10.3}$$

$$\frac{x}{10.3} = \sin 57°$$

Multiply both sides by 10.3.

$$10.3 \times \frac{x}{10.3} = \sin 57° \times 10.3$$

$$x = 10.3 \sin 57°$$

$$= 8.6383 \ldots$$

$$\approx 8.6 \text{ m}$$

Dreamtime.com/Ramon Grasso

EXAMPLE 6

Finding an unknown angle

Calculate the size of θ, correct to the nearest minute.

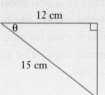

Solution

With reference to θ, 12 cm is the adjacent side. 15 cm is the hypotenuse. The adjacent side and the hypotenuse are in the cos formula. Put the values in the formula, then use the 'undo the cos' buttons on a calculator:

SHIFT COS 12 ÷ 15 = ° ' ".

$$\cos \theta = \frac{12}{15}$$
$$\theta = 36.8698 \ldots^\circ$$
$$= 36° \ 52' \ 11.63"$$
$$\approx 36° \ 52'$$

Exercise 9.04   Sine and cosine

Example 5

1 Use the sin ratio to calculate the value of each pronumeral, correct to one decimal place.

a

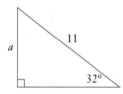

b

c

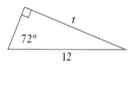

2 Use the sin ratio find the value of each pronumeral, correct to the nearest minute.

a

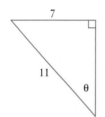

b

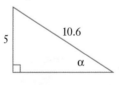

c

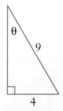

ISBN 9780170413534

3 Use the cos ratio to determine the value of each pronumeral, correct to the nearest metre.

a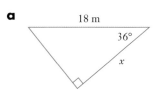
18 m
36°
x

b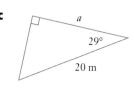
k
72°
25 m

c
a
29°
20 m

4 Calculate the value of each pronumeral using the cos ratio. Express each answer correct to the nearest degree.

Example 6

a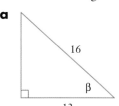
16
β
12

b
8
α
17

c
10 θ
4.5

5 For each triangle, choose the appropriate trig ratio, then calculate the value of the pronumeral, correct to one decimal place.

> When the side you know is the hypotenuse, you will need to use sin if the side with the pronumeral is opposite and cos if it is adjacent.

a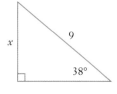
x
9
38°

b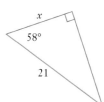
x
58°
21

c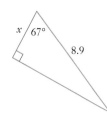
x 67°
8.9

9.05 Sine, cos or tan?

One way to remember the trig ratio formulas is with the phrase 'Only half an hour of algebra' and the trigonometry keys on your calculator in the following order:

Only half	an hour	of algebra
$\dfrac{\text{opposite}}{\text{hypotenuse}}$	$\dfrac{\text{adjacent}}{\text{hypotenuse}}$	$\dfrac{\text{opposite}}{\text{adjacent}}$
sin	cos	tan

Some students like to say SOH-CAH-TOA to help them remember the formulas. Choose the memory method that works for you.

EXAMPLE 7

Which trigonometry ratio can we use to find the value of the pronumeral in each triangle?

a

b

Solution

a   The marked sides are the adjacent,   Use the cos ratio.
a, and the hypotenuse, h. 'an hour' is
the middle section of the phrase, and
the middle trig key on the calculator
is cos .

b   The marked sides are the opposite, o,   Use the tan ratio.
and the adjacent, a. 'of algebra' is the
last section of the phrase, and the last
trig key on the calculator is tan .

Exercise 9.05 Sin, cos or tan?

Example
7

1 In each triangle, which trig ratio could be used to find the value of the pronumeral?

a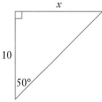

b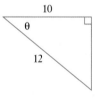

c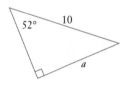

2 Find the value of each pronumeral, correct to one decimal place.

a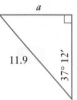

b

c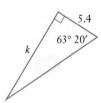

3 Calculate the size of the marked angle, correct to the nearest minute.

a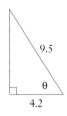
9.5
θ
4.2

b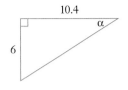
10.4
α
6

c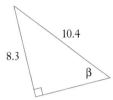
10.4
8.3
β

4 Calculate the size of each angle, correct to the nearest degree.

a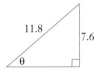
11.8
7.6
θ

b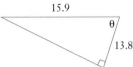
15.9
θ
13.8

c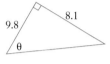
9.8
8.1
θ

5 A 25 m long skateboard ramp is at 17° to the horizontal. How high is the end of the ramp above the ground, correct to one decimal place?

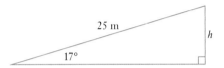
25 m
h
17°

6 Find the value of each pronumeral, correct to one decimal place.

a
17.4
33° 15′
x

b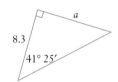
a
8.3
41° 25′

c
67° 11′
8.2
y

7 Determine the size of α, correct to the nearest minute.

a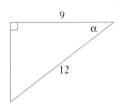
9
α
12

b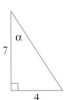
α
7
4

c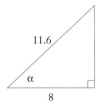
11.6
α
8

9.06 Finding the hypotenuse

EXAMPLE 8

Calculate the length of the hypotenuse, correct to one decimal place.

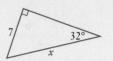

Solution

The sides involved are the opposite and the hypotenuse. Opposite and hypotenuse are in the sin formula, and the opposite side is on the top of the ratio.

$$\frac{7}{x} = \sin 32°$$

Take the reciprocal of both sides to put x on top.

$$\frac{x}{7} = \frac{1}{\sin 32°}$$

Multiply both sides of the equation by 7.

$$\cancel{7} \times \frac{x}{\cancel{7}} = \frac{1}{\sin 32°} \times 7$$

$$= \frac{7}{\sin 32°}$$

The calculator steps to find x are:

$$x \approx 13.2.$$

Exercise 9.06 Finding the hypotenuse

Example
8

1 Calculate the length of the hypotenuse in each triangle. Express your answers correct to one decimal place.

a

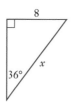

b

c

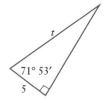

2 Calculate the value of each pronumeral, correct to one decimal place.

a

b

c

3 Scott is designing a wheelchair ramp into a building. The building is 1.5 m higher than the ground. The angle of inclination of the ramp will be 4°.

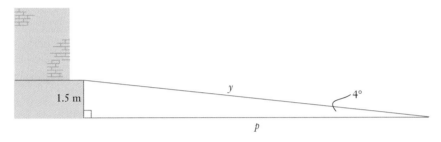

a Calculate the horizontal distance, p, required for the ramp. Express your answer correct to 2 decimal places.

b How long is the ramp, y? Answer correct to one decimal place.

4 Calculate the length of the hypotenuse in each diagram, correct to one decimal place.

a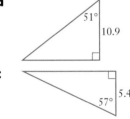

b

c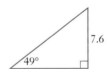

d

5 Determine the value of the pronumeral in each diagram. Express your answers correct to one decimal place.

a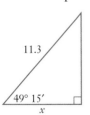

b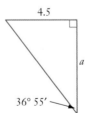

c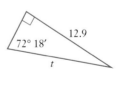

TRIGONOMETRY CROSSWORD

Copy the crossword and use the trigonometry summary below to complete the crossword puzzle.

The word 'trigonometry' comes from two Greek words that combine to mean 'the measurement of triangles'. In modern times, **2**_____ has many applications, including determining the sizes of **6**_____ and **8**_____ in right-angled triangles.

In a **5**____-angled triangle, the side opposite the right angle is called the **7**_____ but the names of the other sides are determined by the reference angle. The **3**_____ side is opposite the reference angle and the **1**_____ side is next to the reference angle.

The ratio of the opposite side to the adjacent side is called the **4**___ _____. We use the sin or the **9**_____ ratio when we do calculations involving the hypotenuse.

SOLUTION TO THE CHAPTER PROBLEM

Problem

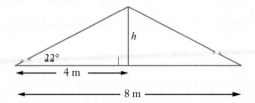

Denis is building a garage. The garage is 8 m wide and the pitch of the roof is 22°. How high does he need to make the truss?

Solution

Height of truss, h, is opposite 22° and the side 4 m is adjacent to 22°. Opposite and adjacent are in the tan formula.

$$\frac{h}{4} = \tan 22°$$

$$h = \tan 22° \times 4$$

$$= 1.6161 \ldots$$

$$\approx 1.62 \text{ m}$$

Denis will need to make the truss 1.62 m high.

- What parts of this chapter did you remember from previous learning?

- What parts of this chapter will you use in your life outside school?

- Are there any parts of this chapter that you don't understand? If yes, ask your teacher for help.

Copy and complete this mind map of the topic, adding detail to its branches and using pictures, symbols and colour where needed. Ask your teacher to check your work.

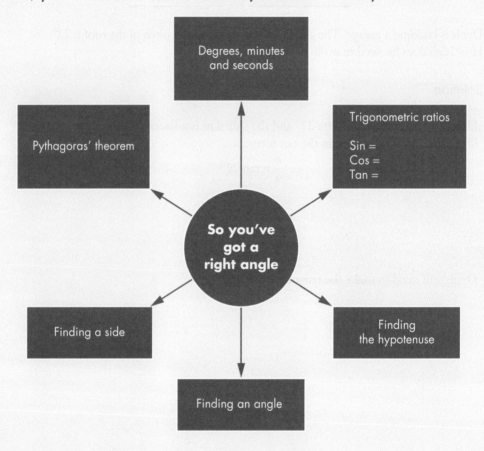

Degrees, minutes and seconds

Trigonometric ratios

Sin =
Cos =
Tan =

Pythagoras' theorem

So you've got a right angle

Finding a side

Finding the hypotenuse

Finding an angle

1 Find the value of each pronumeral, correct to one decimal place.

Exercise
9.01

a

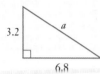

b

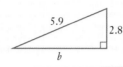

2 Does each triangle contain a 90° angle? Justify your answer.

Exercise
9.01

a

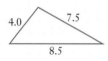

b

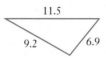

3 Use the tan ratio to find the value of *x*, correct to one decimal place.

Exercise
9.02

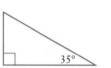

4 Copy and complete each statement.

Exercise
9.03

  a There are _____ minutes in 1 degree.

  b There are _____ seconds in 1 minute.

5 Express each angle correct to the nearest minute.

Exercise
9.03

  a 42° 25' 43"

  b 61° 12' 29"

6 Calculate the size of the angle marked with θ, correct to the nearest minute.

Exercise
9.03

a

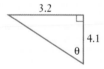

b

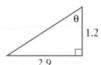

7 Use the sine ratio to find the value of *y*, correct to one decimal place.

Exercise
9.04

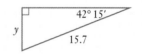

8 Use the cosine ratio to calculate the size of α, correct to the nearest minute.

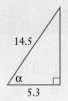

14.5

α

5.3

9 Which ratio, sin, cos or tan, could you use to calculate the value of the pronumeral in each diagram? Determine the ratio, then calculate the value of the pronumeral.

a

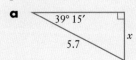

39° 15′

5.7

x

b

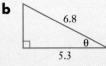

6.8

θ

5.3

Answer to the nearest degree.

c

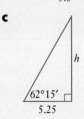

h

62° 15′

5.25

d

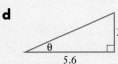

2.7 Answer to the nearest minute.

θ

5.6

10 Calculate the length of the hypotenuse, correct to one decimal place.

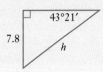

43°21′

7.8

h

Practice set 2

Section A: Multiple-choice questions

For each question, select the correct answer **A**, **B**, **C** or **D**.

1 The cost of sending a parcel weighing 5.5 kg to Malaysia is $88. What is the postage rate?

Exercise 7.01

    A $0.07/kg   **B** $1.60/kg   **C** $16/kg   **D** $484/kg

2 Which graph is a non-linear graph?

Exercise 6.01

    A   **B**   **C**   **D**

3 What is the length of the side marked d, to the nearest kilometre?

Exercise 9.01

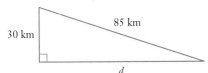

    A   55 km   **B**   80 km   **C**   90 km   **D**   6325 km

4 Which network has a vertex of degree 4 and is able to be traversed?

Exercise 8.01

    A   **B**   **C**   **D**

5 A kangaroo travels 1.5 km in 15 minutes. What is its average speed?

Exercise 7.02

    A   6 km/h   **B**   10 km/h   **C**   16.5 km/h   **D**   22.5 km/h

6 A population of bacteria starts at 200 and is doubling every day. Which graph represents this type of population growth?

Exercise 6.04

    A   **B**   **C**   **D**

Exercise
8.01

7 What is the shortest route from A to B?

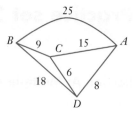

A A to B

B A to D to B

C A to C to B

D A to D to C to B

Exercise
9.03

8 Find θ to the nearest degree if $\tan \theta = 0.859$.

A $31°$ **B** $40°$ **C** $41°$ **D** $59°$

Exercise
6.05

9 A cup of tea is placed on a kitchen bench to cool. At first, it loses heat quickly, but, as time passes, it loses heat more slowly until it reaches room temperature. Which graph best illustrates this?

A
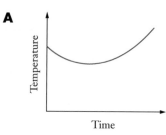

B
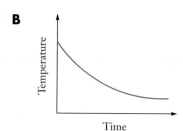

C
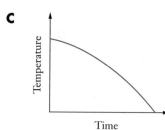

D
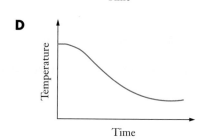

Exercise
7.04

10 Cat food is sold in four different packages:

Pack 1: 24×250 g cans for $39.99

Pack 2: 12×300 g cans for $24.99

Pack 3: 9×400 g cans for $25.99

Pack 4: 6×1 kg cans for $38.50

Which pack is the cheapest price per 100 g?

A   Pack 1   **B**   Pack 2   **C**   Pack 3   **D**   Pack 4

11 Find the length of side *h*.

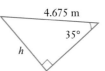

Exercise
9.04

A 2.68 m	**B** 3.27 m
C 3.83 m	**D** 8.15 m

12 Which diagram shows a spanning tree for this network?

Exercise
8.02

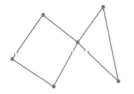

A

B

C

D

Section B: Short-answer questions

1 A catapult is a medieval weapon for hurling large stones. This table gives the height, *h* metres, of a stone thrown by this catapult, *d* metres horizontally from the catapult.

Exercise
6.01

Horizontal distance from catapult (*d* metres)	0	1	2	3	4	5
Height (*h* metres)	5	6.5	7	6.5	5	2.5

   a   Draw the graph of this data and join the points with a smooth curve.

   b   What is the greatest height reached by the stone?

2 A tap drips water at a rate of 18 mL/h. How much water does the tap drip in one week?

Exercise
7.05

Exercise
9.01

3 A firefighter places a ladder on a window sill 4.5 m above the ground. The foot of the ladder is 1.6 m from the wall.

   a   Draw a diagram to show this information.

   b   How long is the ladder? Answer correct to one decimal place.

Exercise
8.03

4 This table shows the distances, in kilometres, by road between 5 cities and towns.

	Crookwell		
Goulburn	48	**Goulburn**	
Canberra	------	90	**Canberra**
Yass	------	85	58
Batemans Bay	------	143	158

   a   Copy and complete this network diagram for these places using the information in the table.

                              • Crookwell

                Yass•

                         • Goulburn

                  •
               Canberra

                          • Batemans Bay

   b   How far is it from Crookwell to Canberra via Goulburn?

   c   Describe 3 possible routes from Yass to Batemans Bay.

   d   Which is the shortest route from Yass to Batemans Bay and how far is it?

Exercise
9.02

5 Find the value of x, correct to 2 decimal places.

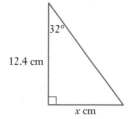

32°

12.4 cm

x cm

Exercise
7.02

6 A tour bus travels at 75 km/h for 5 hours. After a brief stop for refreshments, the bus travels a further distance of 200 km in 4 hours. Find the average speed of the bus over the 9-hour journey. Answer correct to one decimal place.

7 Janine decided to buy a car priced at $14 990. The car depreciates at a rate of 15% p.a.

Exercise 6.03

a Explain why the salvage value can be calculated with the formula $S = 14\,990(0.85)^n$.

b Use this formula to copy and complete this table.

Year (n)	0	1	2	3	4	5	6
Salvage value ($)	14 990						

c Draw the graph for this table of values.

d Use your graph to estimate how long it takes for the car to halve in value; that is, to be worth $7495.

8 Draw a directed network for putting on your shoes and socks in the morning.

Exercise 8.04

9 Calculate the size of θ, correct to the nearest degree.

Exercise 9.05

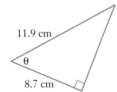

10 A new strain of measles is spreading through the community and the number of people infected with measles is increasing by 7% per day. At the moment, 30 people in the community are infected. The equation for calculating the number of infected people, I, after t days is given by:

Exercise 6.04

$$I = 30 \times (1.07)^t$$

a Copy and complete this table, correct to the nearest whole number.

Number of days (t)	0	1	2	3	4	5	6	7	8	9	10
Number of infected people (I)	30										59

b Draw the graph from this table of values.

c Use your graph to estimate how long it will take before 40 people are infected with measles.

d Do you think the number of people infected with measles can continue to increase exponentially? Give a reason for your answer.

11 Simon the salesman has 8 stores to visit on his rounds. This network shows the distances in kilometres between the stores.

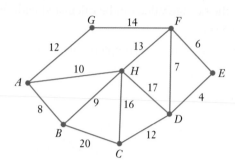

   a   Construct a minimum spanning tree for this network using Prim's algorithm.

   b   What is the minimum distance that Simon can travel in visiting all 8 stores?

12 Judy's Hyundai i30 uses diesel fuel. It has a fuel consumption rate of 7.4 L/100 km.

   a   How far can Judy travel on 50 litres of fuel? Answer to the nearest 5 km.

   b   Diesel fuel costs 119.9c/L. How much does it cost Judy to put 50 L of fuel into her car?

13 This graph shows the height of water in a cylindrical water tank over one month. Describe what the graph is showing.

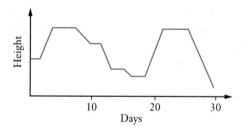

14 A large vat is being filled at a rate of 35 mL/min. Convert this rate to L/day.

15 Gina inspects rural properties to check weed numbers. She has 7 properties to visit and this network shows the distances in kilometres between the properties.

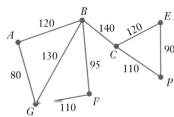

a Use Kruskal's algorithm to construct a minimum spanning tree for this network.

b What is the minimum distance Gina can travel to visit all 7 properties?

c What other factors might affect the order in which Gina chooses to visit the properties?

16 Calculate the length of *h* in this diagram, correct to one decimal place.

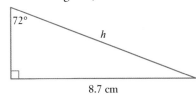

10.

HEALTHY HEART

Chapter problem

Hannah's resting heart rate is 49 beats per minute and she is 17 years old. What is her minimum and maximum target heart rate for aerobic training?

WHAT WILL WE DO IN THIS CHAPTER?

- Calculate and compare heart rates
- Calculate maximum heart rates and target heart rates
- Measure, compare and graph blood pressures
- Interpret situations and solve problems involving heart rates and blood pressure

HOW ARE WE EVER GOING TO USE THIS?

- When exercising at an appropriate level for health and fitness
- When monitoring blood pressure as part of a healthy lifestyle
- To live a longer, healthier life!

10.01 Heart rates

The heart is a large muscle in our chest that pumps blood around our body. Our hearts work very hard for our whole life, from before we're born until we die. If our hearts stop beating for more than a few minutes, we die. If we want to live long and healthy lives, it is essential that we look after our heart.

$$\text{Heart rate in beats per minute} = \frac{\text{number of beats}}{\text{time in seconds}} \times 60$$

EXAMPLE 1

Darren takes his pulse during exercise and counts 52 beats in 20 seconds. Calculate his heart rate in beats per minute.

Solution

Use the heart rate formula.

$$\text{Heart rate} = \frac{52 \text{ beats}}{20 \text{ s}} \times 60$$
$$= 156 \text{ beats/minute}$$

Write the answer.

Darren's heart rate is 156 beats/minute.

EXAMPLE 2

Emma exercised at an intensity that caused a pulse rate of 140 beats/minute. How many times did her heart beat during her 50 minute workout?

Solution

To find beats, multiply by the rate.

$$\text{Number of beats} = 140 \times 50$$
$$= 7000$$

Write the answer.

Emma had 7000 heart beats.

Exercise 10.01 Heart rates

1 When Jessica was jogging, she checked her pulse. She counted 39 beats in 20 seconds. Calculate Jessica's pulse in beats/minute.

2 Calculate the pulse rate in beats per minute for these 6 people on a fun run.

		Number of heart beats	Time in seconds	Pulse in beats/minute
a	James	63	30	
b	Kim	15	10	
c	Abduh	37	15	
d	Xander	51	20	
e	Jabir	65	25	
f	Fenella	78	30	

3 During a 40-minute fun run, Brigit's average heart rate was 123 beats/minute. How many times did her heart beat during the fun run?

4 During a panic attack, John's heart was beating at a rate of 42 beats/15 seconds.

   a Calculate John's heart rate during the panic attack in beats/minute.

   b John's normal, resting pulse is 76 beats/minute. During the 6-minute panic attack, how many more times did his heart beat than while he was resting?

5 This table shows normal pulse rates for healthy people of different ages.

Age	Pulse rate in beats/minute
3 to 4 years	80–120
5 to 6 years	75–114
7 to 9 years	70–110
10 years or more, including adults and seniors	60–100

   a How many times does a healthy 5-year-old's heart beat in 20 seconds?

   b Tommy is a healthy young boy. His heart rate is 29 beats in 15 seconds. Approximately how old is Tommy?

   c Bella's teacher sent her to the sick bay. The school nurse measured Bella's heart rate at 19 beats in 20 seconds. Bella is 8 years old. Does Bella's pulse indicate that she could be sick? Use a calculation to justify your answer.

6 Before a baby is born, it is called a fetus. This table shows some fetal heart rates. The value for 8 weeks is missing.

Fetal age (weeks)	Average heart rate in beats/minute
5	80
6	103
7	126
8	
9	175
11	150
12–41	140

a From the age of 5 weeks to 9 weeks, fetal heart rate increases at a rate of approximately 3.3 beats/minute per day. Calculate an approximate average heart rate for an 8-week-old fetus.

b A fetus' heart rate can be used to estimate its age. How many days old is a fetus whose heart beats 27 times in 15 seconds?

c Use a spreadsheet or pen and paper to construct a line graph to show fetal heart rates at different fetal ages. Put age in weeks on the horizontal axis and heart rate on the vertical axis.

7 A full-term pregnancy is 41 weeks long. If a baby is born before 41 weeks, it is called premature or pre-term. Heart rates of pre-term and full-term babies are different during labor.

	Average heart rate during labor
Full-term baby	140 beats/minute
Premature baby	155 beats/minute

During his birth, baby Dillan's heart beat 39 times every 15 seconds. Is Dillan a full-term or a premature baby? Justify your answer.

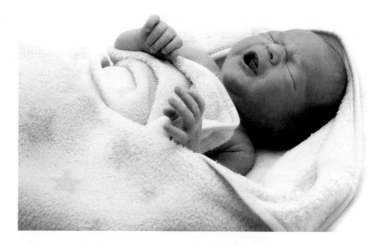

Shutterstock.com/Anna Grigorjeva

PRACTICAL ACTIVITY

RESTING VS EXERCISE PULSE

For this activity, you will need to work in groups with a stopwatch and access to some stairs.

1 Use YouTube to search for a video on how to take your pulse on your wrist.

2 Measure your resting pulse by counting the number of times your heart beats in 15 seconds.

3 Convert your resting pulse to a rate measured in beats/minute.

4 Exercise for 2 minutes by walking up and down stairs.

5 Measure your exercise pulse in beats/minute.

6 Copy and complete the table for members of your group by measuring your pulse rate for 1 minute, 2 minutes, 5 minutes and 10 minutes after exercise.

YouTube

Name	Resting pulse	Exercise pulse	Pulse after exercise			
			1 minute	2 minutes	5 minutes	10 minutes

7 Graph each person's exercise pulse and their pulse for up to 10 minutes after exercise.

8 Who had the lowest resting pulse? Who had the highest?

9 How does a resting pulse change during 2 minutes of stair-climbing?

10 Whose pulse changed the most during the stair-climbing exercise?

11 Whose pulse was the quickest to return to the resting rate after exercise?

12 Some people claim that the lower a person's resting pulse is, the quicker the exercise pulse returns to a resting pulse. Does the data support this claim? Why or why not?

13 Resting pulse rates in beats/minute are one measure of health. This table shows a resting-pulse health rating for 16 to 25-year-olds.

Excellent	Good	Concern	Worrying
55 to 60	60 to 65	66 to 80	More than 80

Assess your current health according to the data in this table.

10.02 Heart rate formulas

When we exercise, our heart beats faster than normal. After we finish exercising, the time it takes for our heart to return to its resting rate is a measure of our fitness. In this section, we will look at formulas for maximum and target heart rates.

Maximum heart rates (MHR)

MHR for males = 220 – age in years

MHR for females = 226 – age in years

EXAMPLE 3

Brittany is 16 years old. What is her maximum heart rate?

Solution

Use the 'maximum heart rate for females' formula.

$$\text{MHR} = 226 - 16$$
$$= 210 \text{ beats/min}$$

Target heart rate

Our **target heart rate** (**THR**) or **training heart rate** is the heart rate we want to achieve during aerobic (or cardiovascular) training to provide maximum health for our heart and lungs. Its value can be calculated using a formula that depends on resting heart rate (RHR), maximum heart rate (MHR) and intensity of exercise (*I*).

Target heart rate

$\text{THR} = I \times (\text{MHR} - \text{RHR}) + \text{RHR}$ where:

RHR = resting heart rate

MHR = maximum heart rate

   I = exercise intensity as a decimal.

Target heart
rates

EXAMPLE 4

This table shows the exercise intensities (I) for different outcomes, expressed as percentages.

Recovery or weight loss	Aerobic	Anaerobic
60% to 70%	70% to 80%	80% to 90%

Ali works out in the gym every day doing aerobic exercise. He is 19 years old with a resting pulse of 62 beats/minute. What is his minimum and maximum target heart rate for aerobic exercise?

Solution

Use the formula for male maximum heart rate to calculate Ali's MHR.

$$MHR = 220 - 19$$
$$= 201 \text{ beats/min}$$

RHR = 62 beats/min
$I = 0.7$ (70%) to 0.8 (80%) for aerobic exercise.

Use the target heart rate formula to calculate the minimum rate.

$$\text{Minimum THR} = 0.7 \times (201 - 62) + 62$$
$$= 159.3 \text{ beats/min}$$

Use the target heart rate formula to calculate the maximum rate.

$$\text{Maximum THR} = 0.8 \times (201 - 62) + 62$$
$$= 173.2 \text{ beats/min}$$

Write the answer.

For aerobic exercise, Ali should aim for a target heart rate from 159 to 173 beats/min.

Exercise 10.02 Heart rate formulas

The questions refer to this table showing 6 people of different ages and their resting heart rates.

	Name	Sex	Age	Resting pulse
a	Jordan	Male	15	65
b	Olivia	Female	18	55
c	Hunter	Male	20	73
d	Katherine	Female	17	50
e	Hassin	Male	21	59
f	Aisha	Female	19	68

Example 3

1 Calculate the maximum heart rate for each person in the table.

2 While we are asleep, our heart is at its resting pulse. Olivia sleeps for 7 hours each night.

   a How many minutes are in 7 hours?

   b How many times does Olivia's heart beat while she is asleep?

3 How many times does Hassin's heart beat during a 30-minute nap?

4 When Jordan is training in the pool, he keeps his heart rate at 135 beats/minute. His training sessions are 45 minutes long. How many times does his heart beat during a training session?

5 17-year-old Katherine goes to the gym each day and exercises at a pulse rate of 80% of her maximum heart rate.

   a What is Katherine's maximum heart rate?

   b Calculate Katherine's exercise pulse rate.

6 Mark's maximum heart rate is 197 beats per minute. How old is he?

7 Megan's maximum heart rate is 215 beats per minute. How old is she?

Example 4

8 The 6 people in the table on the previous page go to the gym for different types of training, as shown below. Calculate the minimum and maximum target heart rate for each person. Calculations for Jordan are shown as an example.

Recovery or weight loss	Aerobic	Anaerobic
60% to 70%	70% to 80%	80% to 90%

	Name	Training	Minimum target heart rate	Maximum target heart rate
a	Jordan	Anaerobic	$0.8 \times (205 - 65) + 65 = 177$	$0.9 \times (205 - 65) + 65 = 191$
b	Olivia	Aerobic		
c	Hunter	Weight loss		
d	Katherine	Anaerobic		
e	Hassin	Aerobic		
f	Aisha	Weight loss		

9 When we're resting, our heart pumps 70 mL of blood every time it beats.

   a How many litres of blood does Jordan's heart pump every minute when he is resting?

   b When she is exercising vigorously, Katherine's pulse is 180 beats per minute and her heart pumps 21 L of blood every minute. How many mL of blood does Katherine's heart pump with each beat when she is exercising vigorously?

ISBN 9780170413534

PULSE RATES DURING EXERCISE

In this activity, you will construct a spreadsheet to show target heart rate information, suitable for displaying in a gym.

1 Ask your teacher to download this 'Pulse rates during exercise' spreadsheet from NelsonNet. Enter 72 in cell C2, an average resting pulse.

Pulse rates
during
exercise

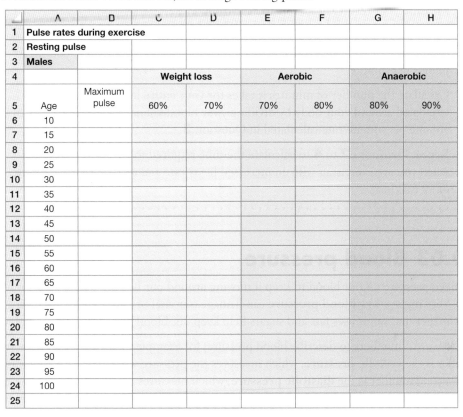

	A	B	C	D	E	F	G	H
1	Pulse rates during exercise							
2	Resting pulse							
3	Males							
4			Weight loss		Aerobic		Anaerobic	
5	Age	Maximum pulse	60%	70%	70%	80%	80%	90%
6	10							
7	15							
8	20							
9	25							
10	30							
11	35							
12	40							
13	45							
14	50							
15	55							
16	60							
17	65							
18	70							
19	75							
20	80							
21	85							
22	90							
23	95							
24	100							
25								

2 For males, maximum pulse rate is 220 – age in years. In cell B6, type the formula **=220-A6** and then press enter. If you have entered the formula correctly, 210 will appear in cell B6.

3 Click on cell B6, then **Fill down** to cell B24.

4 In cell C6, enter the formula **=0.6*(B6 – \$C\$2) + \$C\$2**, then **Fill down** to cell C24.

5 Enter these formulas for cells D6 to H6.

D6	E6	F6	G6	H6
=0.7*(B6 – C2) + C2	=D6	=0.8*(B6 – C2) + C2	=F6	=0.9*(B6 – C2) + C2

Fill down all the formulas to row 24 to create a table of target heart rates.

6 Change the resting pulse at cell C2 to see the table's values change. Then construct a similar spreadsheet for females, using a different formula for maximum pulse.

INVESTIGATION

HOW LONG DOES IT TAKE YOUR HEART TO CIRCULATE ALL OF YOUR BLOOD?

1 Search the Internet for a 'blood volume calculator' or use the one at the **Easy Calculation** website to calculate the volume of blood in your body based on your gender, weight and height. Write your blood volume.

2 Measure your resting heart rate in beats/min.

3 The typical heart pumps 70 mL = 0.07 L of blood during each beat. Multiply your resting heart rate by 0.07 L to calculate the number of litres of blood your heart pumps per minute.

4 Calculate the number of minutes it takes for your entire blood volume to be circulated around your body by dividing your blood volume from Question **1** by the volume per minute rate calculated in Question **3**.

5 Repeat Questions **2** to **4** to calculate the time required to circulate your blood when you are exercising with an increased heart rate and an increased pump rate of 90 mL/beat.

10.03 Blood pressure

We need pressure on our blood to keep it moving around our body to deliver oxygen to our cells. Very high or low **blood pressure** can damage our vital organs. Regular exercise and a healthy diet that minimises fats and sugars help keep our blood pressure at a healthy level.

Doctors use 2 numbers to describe blood pressure; for example, '130 over 85'. The first number is the **systolic pressure**, which measures the pressure when our heart is pumping. The second number is the **diastolic pressure**, which measures the pressure when the heart is relaxed. Blood pressure is measured in mm of mercury, written as 'mmHg'.

Categories of blood pressure

Normal blood pressure: below 120/80

High blood pressure (hypertension): 140/90 or higher.

Low blood pressure (hypotension): Less than 90/60 with symptoms of light-headedness.

> 'hyper' means high
> 'hypo' means low
> It's easy to remember which one is which.
> Hypo has an o in it, just like low.

Elite athletes can have blood pressure at 90/60 without it being considered low.

EXAMPLE 5

Javaneh's blood pressure is 116/74.

a What is her systolic pressure?

b What is the pressure in Javaneh's arteries when her heart isn't pumping?

Solution

The first number is the systolic pressure.

Javaneh's systolic pressure is 116 mmHg.

The second number is diastolic pressure, the pressure when the heart is relaxed.

When her heart isn't pumping, the pressure in Javaneh's arteries is 74 mmHg.

Exercise 10.03 Blood pressure

1 Ben's blood pressure reading is 125/72.

a What is Ben's diastolic blood pressure reading?

b What is the pressure in Ben's arteries when his heart is pumping?

2 When she was standing, Judy's blood pressure was 138/87. Two minutes after she sat down, her blood pressure was 120/75. By how much did Judy's systolic blood pressure decrease when she sat down?

3 This chart shows blood pressure categories.

Systolic (mmHg)		Diastolic (mmHg)	Evaluation
120 or less	and	80 or less	Normal blood pressure
121–139	or	81–89	Pre-hypertension (Pre-high blood pressure)
140–159	or	90–99	Stage 1 high blood pressure
160 or higher	or	100 or higher	Stage 2 high blood pressure
Higher than 180	or	Higher than 110	Emergency situation

a Darren's blood pressure is 120/85. Does he have normal blood pressure?

b Sylvia's blood pressure is 123/70. Explain why her doctor said she has pre-high blood pressure.

c Morgan is 25 weeks pregnant and she has developed gestational hypertension (high blood pressure in pregnancy). Her blood pressure is 145/95. If she wasn't pregnant, how would her high blood pressure be classified?

High blood pressure in pregnancy is very dangerous. It puts both the mother's and the baby's life in danger. Untreated gestational hypertension leads to eclampsia (convulsions in a pregnant woman suffering from high blood pressure), which is usually fatal for the mother.

d Mandy doesn't feel very well. She is finding it hard to concentrate and she has a headache. When she took her blood pressure on her home monitor, it was 195/108. What do you recommend Mandy do?

4 This table shows normal, resting blood pressure readings for children.

Age	Systolic blood pressure	Diastolic blood pressure
Birth (12 hours)	60–76	31–45
Neonate (96 hours)	67–84	35–53
Infant (1–12 months)	72–104	37–56
Toddler (1–2 years)	86–106	42–63
Preschooler (3–5 years)	89–112	46–72
School age (6–9 years)	97–115	57–76
Preadolescent (10–11 years)	102–120	61–80
Adolescent (12–15 years)	110–131	64–83

Source: eMedicineHealth.com

a Rachel is 8 years old and her blood pressure is 98/60. Does she have normal blood pressure?

b Approximately how old is a healthy child whose blood pressure is 90/70?

c By approximately what percentage does a baby's blood pressure increase from when it is 12 hours to when it is 96 hours old?

d Describe how a child's blood pressure changes with age.

Mayo Clinic

PRACTICAL ACTIVITY

BLOOD PRESSURE AFTER EXERCISE

For this group activity, you need an automatic blood pressure-measuring machine and access to a walking area.

1 Search the **Mayo Clinic** website or YouTube for a video 'How to measure blood pressure using an automatic monitor.'

2 Measure each group member's resting blood pressure.

3 Go for a brisk group walk for 5 minutes, then measure everyone's blood pressure again.

4 Construct a scatterplot showing each person's resting blood pressure against their blood pressure after their walk.

5 What happens to blood pressure after exercise?

6 Does exercise change everybody's blood pressure in the same way?

7 Describe the resting blood pressure of individuals whose blood pressure changed the most during exercise.

10.04 Blood pressure graphs

Blood pressure isn't constant. Everyone's blood pressure fluctuates, but it should remain within the shaded bands on the graph below. When you are asleep, your blood pressure is 10% to 15% lower than normal. Blood pressure graphs allow health professionals to obtain a lot of information quickly. Some professionals use 2 lines to graph blood pressure, whereas others join the systolic and diastolic readings with a vertical line.

Graphing blood pressure

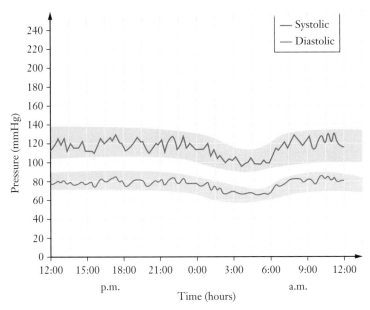

Source: eMedicineHealth.com

EXAMPLE 6

When Jenny was 9 weeks pregnant, she had a miscarriage and lost a lot of blood. At time 1315, the hospital staff began treating her low blood volume and monitoring her blood pressure. The table shows her blood pressure.

Time	1315	1320	1325	1330	1335	1340	1345	1350	1355
Systolic pressure (mmHg)	70	75	80	90	90	95	95	100	105
Diastolic pressure (mmHg)	55	60	60	65	65	70	75	80	80

Graph Jenny's blood pressure readings.

Solution

Draw a cross in the positions for the systolic and diastolic readings, then join the crosses with a vertical line.

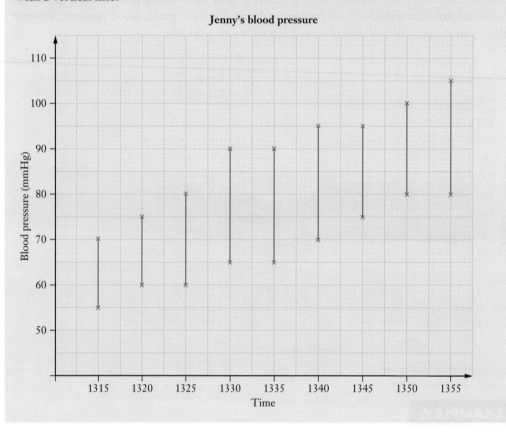

Jenny's blood pressure

Exercise 10.04 Blood pressure graphs

Example 6

1 During an operation, medical staff discovered that Ian was allergic to the anaesthetic they were using. The table shows what happened to his blood pressure.

Time	11 a.m.	1 p.m.	3 p.m.	5 p.m.	7 p.m.	9 p.m.	11 p.m.	1 a.m.	3 a.m.	5 a.m.
Systolic pressure (mmHg)	140	180	180	160	170	180	200	175	150	140
Diastolic pressure (mmHg)	90	130	135	130	130	145	150	135	120	105

Construct a graph of Ian's blood pressure.

ISBN 9780170413534

2 This graph shows normal blood pressure ranges at different ages.

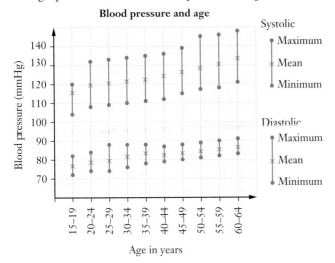

Blood pressure and age

Age in years

a Normal systolic blood pressure for a 55–59-year-old is 118 to 144.
What is the normal diastolic range for a 55–59-year-old?

b What are the mean systolic and diastolic readings for a 55–59-year-old?

c What is the systolic blood pressure range for a 30–34-year-old?

d What are the average systolic and diastolic values for a 22-year-old?

e According to the graph, what should your blood pressure be?

f In any age group, how is the range of systolic blood pressure readings different to
the range of the diastolic readings?

3 Some people have normal blood
pressure, but their blood pressure rises
just because their doctor is measuring it!
This is known as 'white coat
hypertension'. This graph shows a case
of white coat hypertension.

a At what time did the white coat
hypertension occur? Express
your answer in 12-hour time.

b What was the patient's blood
pressure at this time?

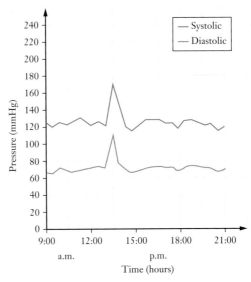

Source: Chameides, Leon, Ricardo A. Samson, Stephen
M. Schexnayder, and Mary Fran Hazinski, eds. Pediatric
Advanced Life Support Provider Manual: Professional
Edition. United States of America: American Heart
Association, 2011. http://www.emedicinehealth.com/
pediatric_vital_signs/article_em.htm

10. Healthy heart

4 Many stroke victims have a blood pressure graph similar to this one.

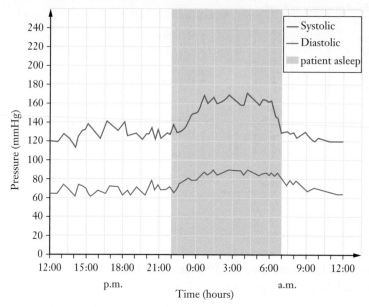

Source: eMedicineHealth.com

a Describe what this graph is showing.

b Why do you think this type of blood pressure is called 'nocturnal hypertension'?

5 This blood pressure graph is typical of individuals who have hardening of their arteries. Hardening of the arteries results from fat and/or plaque build-up that narrows the arteries and makes them inflexible.

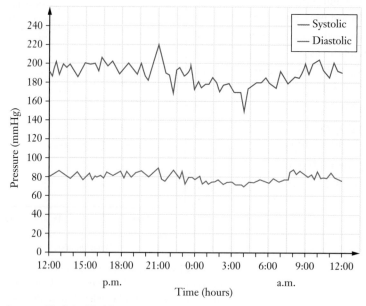

Source: eMedicineHealth.com

ISBN 9780170413534

a At what time was this patient's blood pressure the lowest?

b What was the patient probably doing at the time?

c When was the patient's blood pressure the highest and what was the reading?

d How is the blood pressure of individuals with hardening of the arteries different from that of individuals with normal blood pressure?

e How might this difference be explained?

INVESTIGATION

USING A SPREADSHEET TO GRAPH BLOOD PRESSURE

In this activity, you will create a spreadsheet to record and graph your blood pressure over the next 6 maths lessons.

1 Open a new spreadsheet page and enter the labels as shown below. Enter today's date and the next four dates in column A.

	A	B	C	D
1		Blood pressure records		
2				
3				
4		Systolic	Diastolic	
5	1-Apr			
6	2-Apr			
7	3-Apr			
8	4-Apr			
9	5-Apr			
10	6/			
11				
12				

2 Enter the systolic and diastolic data; for example (130, 67), (138, 82), (130, 70), (133, 78), (146, 88) and (126, 76).

3 Select **Table** from the **Insert** menu, or press CTRL-T.

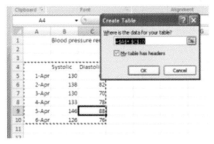

4 Select the data in the table, then select **Insert**, **Graph** and select the top-left line graph to draw a graph.

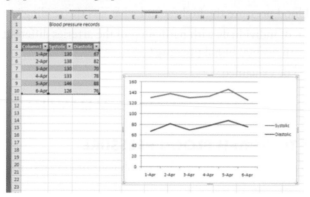

KEYWORD ACTIVITY

BACKWARDS CROSSWORD

Copy the crossword below. The key words for this chapter include:

rate	pulse	diastolic	blood pressure	fetus
arteries	hypotension	heart	systolic	hypertension

Your task is to arrange the words in the crossword, then write the clues.

SOLUTION TO THE CHAPTER PROBLEM

Problem

Hannah's resting heart rate is 49 beats per minute and she is 17 years old. What is her minimum and maximum target heart rate for aerobic training?

Solution

Use the formula for female maximum heart rate to calculate Hannah's MHR.

$$MHR = 226 - 17$$

$$= 209 \text{ beats/min}$$

Use the target heart rate formula to calculate the minimum rate.

$$THR = I \times (MHR - RHR) + RHR$$

$$RHR = 49 \text{ beats/min}$$

$$I = 0.7 \ (70\%) \text{ to } 0.8 \ (80\%) \text{ for aerobic exercise}$$

$$\text{Minimum } THR = 0.7 \times (209 - 49) + 49$$

$$= 161 \text{ beats/min}$$

Use the target heart rate formula to calculate the maximum rate.

$$\text{Maximum } THR = 0.8 \times (209 - 49) + 49$$

$$= 177 \text{ beats/min}$$

For aerobic exercise, Hannah should aim for a target heart rate from 161 to 177 beats/min.

- What part of the chapter did you find the most interesting?

- Are you going to make any changes to your lifestyle as a result of completing this chapter?

- What members of your family could benefit from the content of this chapter?

- Is there any part of this chapter that you don't understand? If yes, ask your teacher for help.

Copy and complete this mind map of the topic, adding detail to its branches and using pictures, symbols and colour where needed. Ask your teacher to check your work.

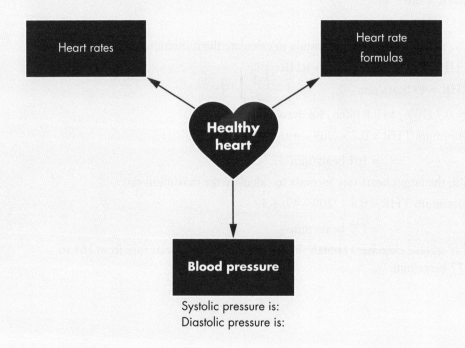

1 Allyson counted 32 heart beats in 24 seconds. What is her pulse rate in beats/minute?

2 During strenuous exercise, Billy's pulse rate was 164 beats/minute. How many times did Billy's heart beat during the 15 minutes of the exercise session?

3 Annie is 17 years old. What is her maximum heart rate?

4 Dae's maximum heart rate is 204 beats/minute. How old is he?

5 Use the formula THR = $I \times$ (MHR − RHR) + RHR to calculate Nate's target heart rate when he wants to exercise at an intensity of 0.85. Nate is 17 years old and his resting heart rate is 66 beats/minute.

6 Emily's blood pressure is 118/75. What is the pressure in her arteries when her heart is beating?

7 Kasey has white coat syndrome. This graph shows his blood pressure for 24 hours.

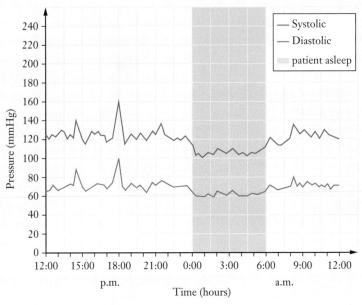

a What does 'white coat syndrome' mean?

b At what time did Kasey have his blood pressure checked by his doctor?

c What was Kasey's blood pressure when his doctor checked it?

d Does Kasey have hypertension?

e During what time was he asleep?

MEASUREMENT

11.

APPLYING TRIGONOMETRY

Chapter problem

Early Australian builders knew that a large roof pitch helped prevent leaks and helped keep houses cool. The early roof pitches were often as large as 40°. Today, most roof pitches are about 22°.

How much higher was the roof of a 16 m wide early Australian building with a 40° roof pitch than a modern building with the same width and a 22° roof pitch?

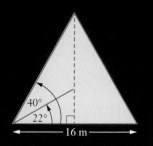

WHAT WILL WE DO IN THIS CHAPTER?

- Use trigonometry to solve practical problems, including those involving angles of elevation and depression
- Describe direction using compass and true bearings, and use trigonometry to solve problems involving bearings
- Learn about traditional navigation methods

HOW ARE WE EVER GOING TO US THIS?

- Some sports, for example, orienteering and sailing, use bearings
- When we navigate in bushland or at sea, a knowledge of compass bearings can prevent us from getting lost
- Some workplaces require trigonometry calculations; for example, carpentry and land surveying

Trigonometry
review

11.01 Using the trigonometric ratios

In chapter 9, we used sine, cosine and tangent ratios in right-angled triangles to calculate sides and angles. Let's review these.

EXAMPLE 1

Calculate the length of x, to 2 decimal places.

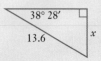

Solution

x is opposite the angle and 13.6 is the hypotenuse. Use the sine ratio.

Multiply both sides by 13.6.

$$\frac{x}{13.6} = \sin 38° \ 28'$$

$$x = 13.6 \sin 38° \ 28'$$

$$\approx 8.46$$

EXAMPLE 2

Calculate α, correct to the nearest minute.

Solution

The opposite side is 14.9 and the adjacent side is 8.5. Use tan.

$$\tan \alpha = \frac{14.9}{8.5}$$

$$\alpha = 60.29654 \ldots$$

$$= 60° \ 17' \ 47.57''$$

$$\approx 60° \ 18'$$

Exercise 11.01 Using the trigonometric ratios

Calculate the size of the pronumeral in each triangle. Express lengths correct to 2 decimal places and angles correct to the nearest minute.

1

2
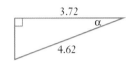

Example
1

Example
2

3
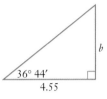

4

5

6

7

8
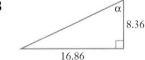

9 The hypotenuse of a right-angled triangle is 12 cm long and one of the angles is 40°. Calculate the lengths of the other 2 sides, correct to one decimal place.

11.02 Trigonometry problems

We can use trigonometry to solve a large variety of practical problems. You might be surprised by some of the applications in this exercise.

There are 3 steps to solving a trigonometry problem.

1 Find a right-angled triangle in the diagram.

2 Locate lengths and angles in the right-angled triangle.

   Disregard everything else in the diagram that you don't need.

3 Choose sin, cos or tan and solve the problem.

EXAMPLE 3

360 m

25°

Shutterstock.com/alexfe

Zac is a keen skier. The angle of inclination of his favourite ski slope is 25° and the altitude of the top of the slope is 360 m higher than the bottom of the slope. What is the length of the slope, correct to the nearest metre?

Solution

In the diagram, locate a right-angled triangle, angles and lengths.

Disregard everything else in the picture.

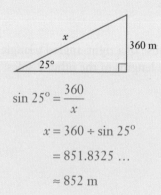

x

360 m

25°

The sides involved are the opposite and the hypotenuse. Use sine.

$$\sin 25° = \frac{360}{x}$$

$$x = 360 \div \sin 25°$$

$$= 851.8325 \ldots$$

$$\approx 852 \text{ m}$$

Write the answer.

The slope is 852 m long.

ISBN 9780170413534

EXAMPLE 4

A fun run course is uphill. During the run, participants cover a horizontal distance of 8 km and their height above sea level increases from 125 m to 394 m. Calculate, correct to the nearest minute, the angle of inclination of the slope.

Solution

Locate a right-angled triangle, angles and lengths.

Disregard everything else in the picture.

Change in height = 394 − 125

$\qquad$ = 269 m

Convert 8 km to 8000 m so all length measurements are in the same units.

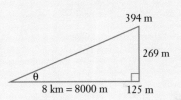

Opposite and adjacent sides are involved. Use the tan ratio.

$$\tan \theta = \frac{269}{8000}$$

$$\theta = 1° \, 55' \, 32.88''$$

$$\approx 1° \, 56'$$

Write the answer.

The angle of inclination of the slope is 1° 56′.

Exercise 11.02 Trigonometry problems

Example
3

1 During a flood, Grant needed to work out the width of a river. Use Grant's measurements to determine the width of the flooded river, correct to the nearest metre.

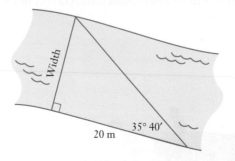

Example
4

2 Kim is taking a shot at goal. What is the largest angle, θ, at which he can kick the ball for it to go into the goal? Answer correct to the nearest minute.

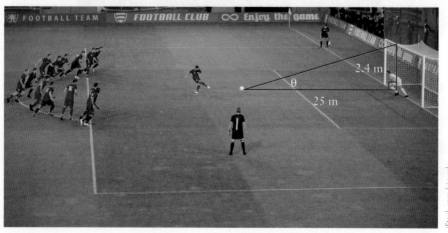

3 A boat's anchor chain is 24 m long, and it is making an angle of 31° with the top of the water. How deep is the water, correct to one decimal place?

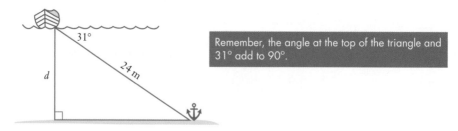

Remember, the angle at the top of the triangle and 31° add to 90°.

4 An advertising blimp is anchored to the ground by a 35 m long rope. The angle between the rope and the ground is 68°. How high is the advertising blimp above the ground? Answer correct to one decimal place.

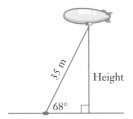

5 Scientists want to calculate the depth of a crater on the Moon. Use this picture and diagram to help you calculate the depth, correct to the nearest metre.

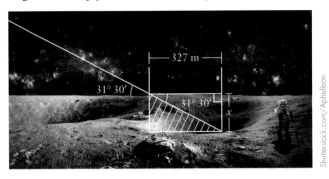

6 At low tide, a wharf is 2.4 m above the water. The ramp joining the wharf to a floating pontoon makes an angle of 12° to the horizontal. Calculate the length of the ramp, correct to one decimal place.

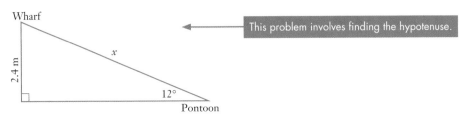

This problem involves finding the hypotenuse.

7 The council is going to build a child's slide in the park. The top of the slide will be 2.6 m high, and the slide will make an angle of 31° with the ground. Calculate the length of the slide, correct to two decimal places.

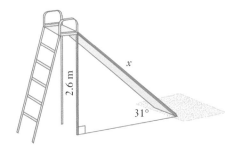

8 The diagram shows a shed roof. Calculate x, the length of each piece of iron sheeting required. Round your answer up to the nearest 10 cm.

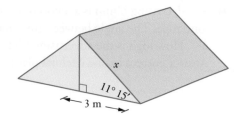

9 Gothic arches are a common feature in the design of Gothic cathedrals. Equilateral triangles are the design basis for most Gothic arches.

a Calculate the size of the angle θ in $\triangle ABC$.

b One Gothic arch is 1.2 m high, so $BC = 1.2$ m.

Calculate the length of AC and hence the width of arch AD correct to 2 decimal places.

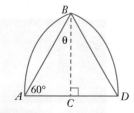

> Remember, the angles in a triangle add to 180°.

10 The communications tower on the slope shown requires a wire support from the top of the tower to the grass. We need to calculate the length of the wire.

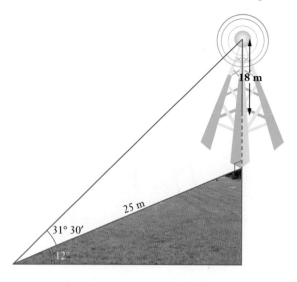

a Use the small triangle to calculate the length of *x* correct to one decimal place.

b Add *x* to 18 to calculate the height of the top of the tower above the baseline, correct to one decimal place.

c Add the 2 parts of the angle together to calculate the size of the base angle of the triangle.

d Use the sine ratio to calculate the length of the hypotenuse that represents the wire. Answer correct to one decimal place.

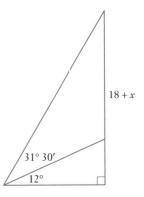

11.03 Angles of elevation and depression

Angles of depression and elevation are used regularly to solve practical measurement problems. The **angle of depression** is the angle the eye turns **down** from the horizontal to look at an object in a lower position. The **angle of elevation** is the angle the eye turns **up** from the horizontal to look at an object in a higher position.

Trigonometry match-up

Angles of elevation and depression

Angles of elevation and depression

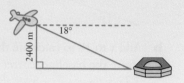

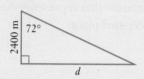

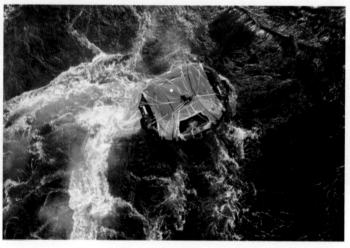

Angle of
depression

Angle of
depression

EXAMPLE 5

A pilot flying at a height of 2400 m saw a life raft in
the sea at an angle of depression of 18°. Calculate
the horizontal distance from the plane to the life raft,
correct to the nearest metre.

Solution

Let d be the horizontal distance from the plane to
the life raft.

The angle of depression and the angle at the top
of the triangle add to 90°.

The top angle $= 90° - 18°$

$\qquad = 72°$

Use tan because the opposite and the adjacent
sides are involved.

$$\tan 72° = \frac{d}{2400}$$

$$d = 2400 \times \tan 72°$$

$$= 7386.4404 \ldots$$

$$\approx 7386 \text{ m}$$

Write the answer.

The horizontal distance from the
plane to the life raft is 7386 m.

Watch out! It's a common error to write the angle of depression as the top angle in the
triangle. Usually, this is wrong, because the angle of depression is outside the triangle.

Alamy Stock Photo/David Wingate

Exercise 11.03 Angles of elevation and depression

In this exercise, express your answers correct to one decimal place (for lengths) or to the nearest degree (for angles), unless instructed otherwise.

1 From the top of a 23 m high mobile phone tower, the angle of depression to the bottom of a set of traffic lights is 11°. How far is the set of traffic lights from the base of the mobile phone tower?

Example
5

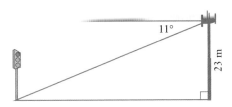

The angle of depression is NOT the top angle in the triangle. Subtract the angle of depression from 90° to determine the size of the top angle.

2 Samantha is flying an ultra-light aircraft at a height of 300 m above the ground. Her angle of depression to the landing strip is 12°. Calculate the horizontal distance from the ultra-light to the landing strip.

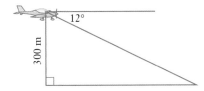

3 A navigation chart shows that the top of a lighthouse is 142 m above sea level. A ship's navigator measured the angle of elevation to the top of the lighthouse as 15°. How far is the ship from the base of the lighthouse?

4 Jill is 45 m away from a tall office block. She measured the angle of elevation to the top of the office block as 74°.

   a Calculate the length of h in metres correct to 3 decimal places.

   b Jill's eyes are 1.55 m above the ground. Calculate the height of the building to one decimal place.

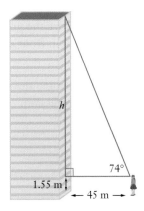

5 When Aamir was 810 m away from Uluru, he measured the angle of elevation to the top of the rock as 20°.

   a Copy and complete the diagram, showing Aamir's measurements on it.

   b Use Aamir's measurements to calculate the approximate height of Uluru. Answer correct to the nearest metre.

6 The third level of the Eiffel Tower in Paris is 276 m above the ground. When Jan was standing on the third level of the tower, she measured the angle of depression to a group of her friends on the ground as 19°. How far were Jan's friends from the base of the Eiffel Tower below Jan?

7 The front of an office block has security lights that come on when they detect movement. The lights are 110 m above the ground, and they light up an 85 m wide strip in front of the building.

   a Calculate the size of angle ω.

   b Determine the size of angle θ, the light's angle of depression.

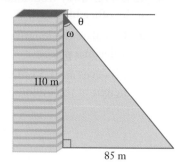

ISBN 9780170413534

8 From a small plane flying at a height of 800 m, the angle of depression to the landing strip is 14°.

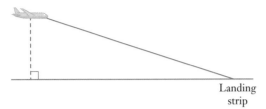

Landing
strip

a Copy the diagram and show the position of the angle of depression.

b Calculate, correct to the nearest metre, the horizontal distance from the plane to the landing strip.

9 During World War II, searchlights and trigonometry were used to calculate the heights of clouds.

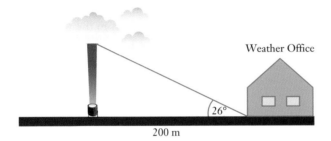

Weather Office

26°

200 m

Calculate the height of the clouds when the angle of elevation to the clouds is 26°.

10 Forensic experts are investigating a murder scene. A man was standing up when he was shot and the bullet went through him 110 cm above the ground. The bullet then lodged in a wall 42 cm behind the man and 23 cm above the ground.

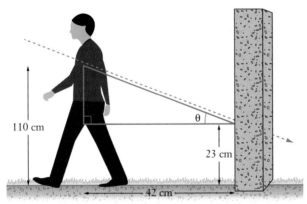

110 cm

θ

23 cm

42 cm

a Copy the triangle and make the adjacent side 42 cm.

b Calculate the length of the opposite side by subtracting 23 from 110.

c Calculate the angle of elevation to the location from where the gun was fired. Answer correct to the nearest minute.

11.04 Compass bearings

The 4 major compass directions are north (N), east (E), south (S) and west (W). The directions halfway between these are northeast (NE), southeast (SE), southwest (SW) and northwest (NW). We use these 8 compass directions to describe wind direction, position of the Sun, shadows, and the path of a plane or ship.

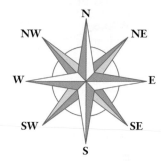

DID YOU KNOW?

It's a common myth that our word 'news' was derived from a combination of the first letters of the directions **n**orth, **e**ast, **w**est and **s**outh. 'News' comes from the word 'new', short for 'new information'. 'New' became 'news' when there was more than one piece of new information.

EXAMPLE 6

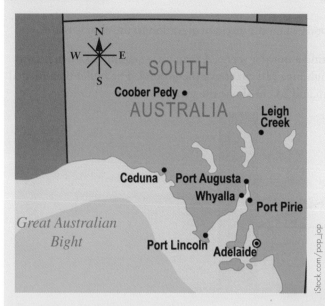

a   What compass direction is it from Coober Pedy to Port Augusta?

b   In what direction is Coober Pedy from Port Augusta?

Solution

a Position the centre of a compass rose on the location the direction is *from*. In this case, the bearing is from Cooper Pedy.

The direction to Port Augusta is in the middle of east and south. The direction from Cooper Pedy to Port Augusta is southeast.

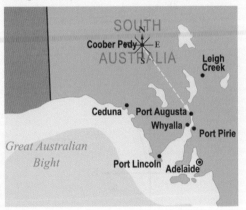

b Now the bearing is from Port Augusta. Position the compass rose on Port Augusta.

The direction to Cooper Pedy is between north and west. Cooper Pedy is northwest of Port Augusta.

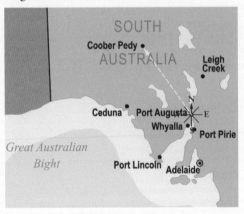

There are many directions that lie between north, south, east and west. If we want to specify a precise direction using a **compass bearing**, we write something like S 15° E, which means 'look south, then turn 15° towards the east'.

When we're using compass bearings, we always start by heading north or south, then give the number of degrees we have to turn towards the east or west.

Compass
bearings

EXAMPLE 7

Write each direction as a compass bearing.

a
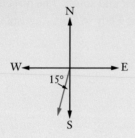

b
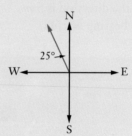

c
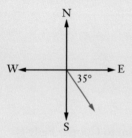

d
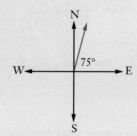

Solution

a   This direction is towards the south, so it starts at S.   S 15° W
    It's 15° towards the west. The bearing ends with 15°W.

b   This direction is initially towards the north with a turn   N 25° W
    of 25° towards the west.

c   Calculate the angle between south and the direction.   S 55° E

    Angle = 90° − 35°
          = 55°

    The direction is towards south with a turn of 55°
    towards the east.

d   The angle between north and the direction = 90° − 75°   N 15° E

                                              = 15°

    The direction is towards north with a turn of 15°
    towards the east.

Exercise 11.04 Compass bearings

1 What is the compass direction:

a from Hobart to Roseberry?

b from Launceston to Bothwell?

c to Roseberry from Queenstown?

d to Launceston from George Town?

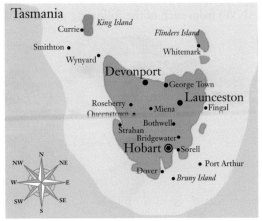

2 In 1986, following an explosion and fire at the Chernobyl nuclear power station, winds carried the radiation cloud over a vast area.

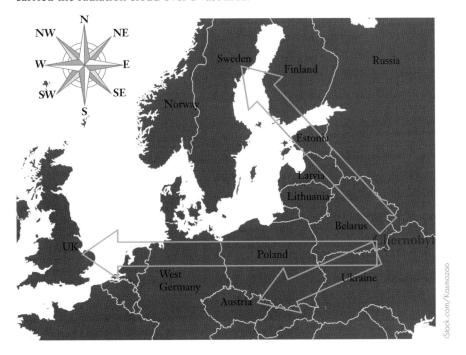

a What compass direction, NW, NE, SE or SW is it from Chernobyl to Austria?

b Easterly winds are winds that come from an easterly direction. What countries experienced radiation clouds as a result of easterly winds?

3 When Muslims pray, they face the central shrine in Mecca's Great Mosque, Islam's holiest place. What is the approximate compass direction (N, S, E, W, NE, SE, SW or NW) from each of the following places to Mecca?

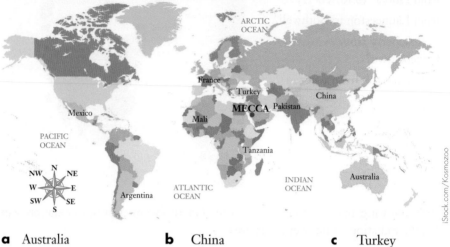

a Australia	**b** China	**c** Turkey
d Pakistan	**e** Mali	**f** Argentina
g France	**h** Tanzania	**i** Mexico

4 Late in the afternoon, the Sun is in the west. Which one of the diagrams shows the correct position of the tree's shadow?

A  **B**  **C**   **D**

Example
7

5 Use compass bearings to describe the direction in each diagram.

a 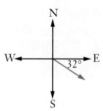 **b**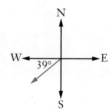

c 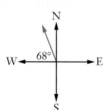 **d**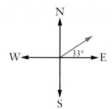

6 A westerly wind brought smoke from a bushfire to Parkes.

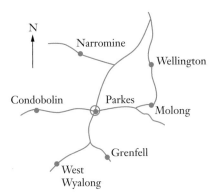

A westerly wind comes from the west.

a Near which town on the map could the bushfire be located?

b If the wind keeps blowing in the same direction, which other town will receive smoke from the fire?

c The wind changed direction and smoke went in the direction of S 10° E. Which town experienced smoke?

7 Winds from over the ocean carry moisture and are likely to bring rain, but winds from over a land mass are likely to be dry. The temperature of the land the winds blow over determines whether the winds will be hot or cold. In winter, winds from over the snowfields will be cold, but summer winds from Central Australia are likely to be dry and hot.

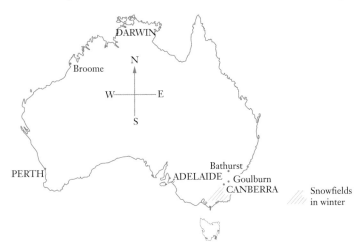

Are the following winds likely to be hot, cold, bring rain or fine weather?

Remember! A wind's direction is the direction it has come *from*.

a westerly winds in Perth

b westerly winds in Bathurst in summer

c northeasterly wind in Goulburn

d southerly winds in Canberra in winter

e southerly winds in Darwin in summer

f southerly winds in Adelaide

g northwesterly winds in Broome

h southeasterly winds in Broome in summer

11.05 True bearings

Bearings

A page of bearings

Identifying bearings

16 points of the compass

NSW map bearings

Bearings match-up

Every which way

Elevations and bearings

Aircraft and ship navigators are responsible for getting their vessels to their destination. They use **true bearings** or **three-figure bearings** (three-digit angles) to specify directions accurately. Three-figure bearings measure angles from true north in a clockwise direction, from 000° to 360°.

On this map of Lord Howe Island, to find the bearing *from* Mt Gower to North Head, we draw a compass rose on Mt Gower, draw a line from Mt Gower to North Head, then measure the angles.

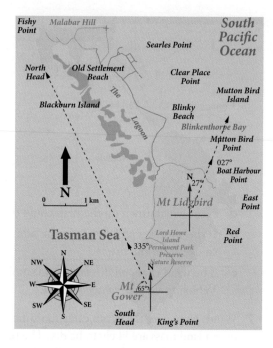

We add all the angles in a clockwise direction from north around to the direction line.

Bearing from Mt Gower to North Head = 90° + 90° + 90° + 65°

$$= 335°$$

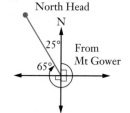

To find the bearing of Mutton Bird Island *from* Mt Lidgbird, position the compass rose on Mt Lidgbird and draw a line from Mt Lidgbird to Mutton Bird Island. The angle between north and the direction line to Mutton Bird Island is 27°, but we write the bearing as 027° because we must use 3 figures.

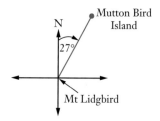

EXAMPLE 8

a What is the bearing of Smuggler's Reef from Town?

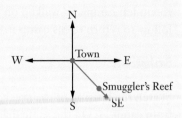

b What is the bearing from Apple Cove to Mt Tims?

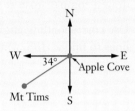

Solution

a The bearing is *from* Town. Position a compass rose on Town and rule the southeast direction.

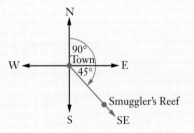

Southeast is exactly in the middle of south and east. $\frac{1}{2}$ of 90° is 45°.

The bearing from Town to Smuggler's Reef = 90° + 45°
= 135°

Add the angles from north, clockwise to the direction of Smuggler's Reef.

b The angle between Mt Tims and south is 90° − 34° = 56°

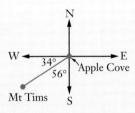

Bearing from Apple Cove to Mt Tims = 90° + 90° + 56°
= 236°

EXAMPLE 9

Kevin left camp and walked on a bearing of 117° to the creek. What bearing should Kevin follow to return to his camp from the creek?

Solution

Sketch the bearing of the creek, B, from camp, A. On the same diagram, draw a compass rose at the creek, B.

Use corresponding angles on parallel lines to find $\angle ABS$, where S is south.

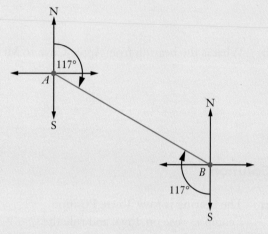

$\angle ABS = 117°$ (corresponding angles, NA || NB)

Add the angles from north to find the bearing back to Kevin's camp from the creek.

$\therefore$ Bearing of camp from creek $= 180° + 117°$

$= 297°$

Exercise 11.05 True bearings

Example 8

1 What is the three-figure (true) bearing and compass bearing from K to H in each diagram?

a

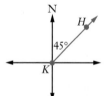

b

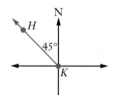

c

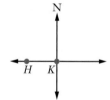

d

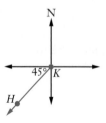

e

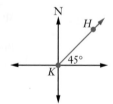

f

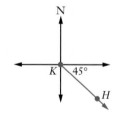

2 What bearing is required to travel from *P* to *T* in each diagram? Answer as a compass bearing and as a true bearing.

a

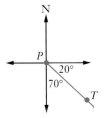

b

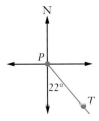

c

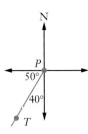

d

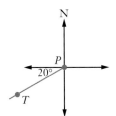

e

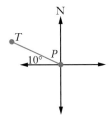

f

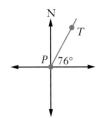

3 Copy and complete the bearings for each trip.

Example 9

	Forward trip bearing	Bearing for the return trip
a	134°	
b	092°	
c		316°
d		065°

4 The bearing of Mecca, Saudi Arabia from Brisbane is 280°.

  a What is the true bearing from Mecca to Brisbane?

  b Express your answer to part **a** as a compass bearing.

5 The direction from Melbourne, Victoria, to Mecca is 279°.

  a What is the true bearing from Mecca to Melbourne?

  b Express your answer to part **a** as a compass bearing.

HOW FAR AND IN WHAT DIRECTION IS MECCA?

In this investigation, you are going to use a website to find the distance and the compass direction from your home to Mecca.

Al Habib

1 Search on the Internet for 'Mecca locator' or go to the **Al Habib** website and select 'Qibla pointer'. Enter the name of your town or city followed by Australia and click 'Find'. Record the distance and the bearing from true north.

For example, from Melbourne, Mecca is 12 741 km and 279°.

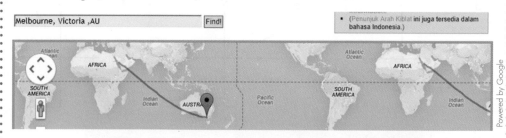

> When we draw straight line compass directions on a map they look curved because a map is a flat representation of the Earth's curved surface.

Magnetic declination

2 A compass gives directions from magnetic north, not true north. The difference between magnetic north and true north is called **magnetic declination**. Magnetic declination depends on where you are. Go to the **Magnetic declination** website and enter the name of your town or city followed by Australia and click **Search map**. Click the link to your town or city and record the magnetic variation and whether the variation is positive or negative.

For example, for Melbourne, the declination is +11°.

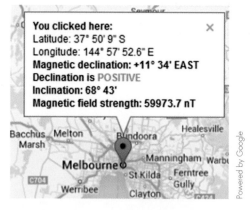

3 The final step is to subtract the size of the magnetic declination from the bearing to Mecca.

True north bearing from Melbourne to Mecca is 279° − 11° = 268°.

11.06 Bearings problems

Bearings

A fishing boat left port and sailed on a bearing of 128° for 30 M (nautical miles) to a fishing reef. How far is the reef east of port? Answer to the nearest nautical mile.

True bearings

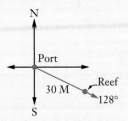

Solution

Construct a right-angled triangle by drawing a line from the reef straight up to the east direction line.

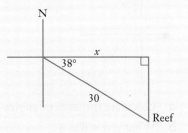

The angle in the triangle = 128° − 90°

$$= 38°$$

Let x be how far the reef is east of port.

Then use trigonometry to find x.

$$\cos 38° = \frac{x}{30}$$

$$x = 30 \cos 38°$$

$$= 23.6403\ldots$$

$$\approx 24 \text{ M}$$

Write your answer.   The reef is 24 M east of port.

Exercise 11.06 Bearings problems

1 A lighthouse is 1500 m from a ship on a bearing of 154°.

   a Explain why $\alpha = 64°$ and $\beta = 26°$ in the diagram.

   b How far is the ship west of the lighthouse?

   c Calculate the distance that the lighthouse is south of the ship.

Example 10

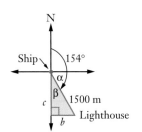

---

2 A light aircraft flew 158 km in a straight line from Maun Airport to Jamo Game Park on a bearing of 119°.

a Calculate the size of the angle α.

b How far is Jamo Game Park east of Maun Airport?

3 Two planes flew out of Cairns to search for a missing yacht. The first plane flew 150 km on a bearing of 065° and the second plane flew on a bearing of 155° for 120 km.

a Calculate the sizes of angles α and β.

b What is α + β?

c Use Pythagoras' theorem to calculate how far the planes are apart.

4 Ben drove along the highway from Parkes to West Wyalong on a bearing of 220°. West Wyalong is 80 km west of Parkes.

a Explain why α = 50°.

b How far, correct to the nearest kilometre, is it along the highway from Parkes to West Wyalong?

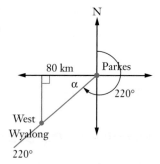

5 Two dive boats sailed out of Broome heading for popular dive sites. The first sailed on a bearing of 325° for 70 nautical miles. The second boat sailed 49 km on a bearing of 235° until it was due south of the first dive boat.

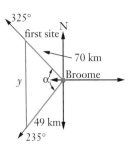

a Explain why $\alpha = 90°$.

b How far, y km, are the two dive sites apart?

Shutterstock.com/Igor Karasi

6 Caleb was looking for sapphires near the city of Emerald in Queensland. He found blue sapphires 60 km on a bearing of 319° from Emerald and yellow sapphires 36 km on a bearing of 229° from Emerald. How far is it between the two locations where Caleb found sapphires?

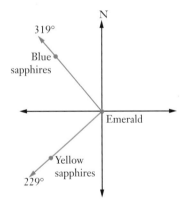

11.07 Traditional navigation

Long before compasses, charts, satellite navigation and modern communications technology, Aboriginals, Torres Strait Islanders and Polynesians made extensive return journeys to distant islands. As you work through the questions in the next exercise you will learn about some of the amazing navigation skills these people learned from their ancestors and passed on to their children.

Imagefolk/Antiquarian Images/Mary Evans

Exercise 11.07 Traditional navigation

1 Some of the words in the following descriptions are missing. Copy the passage and select the words from the table below to complete the blanks.

behind	birds	challenge	clouds
compass	distances	dolphins	east
latitude	navigator	nest	New Zealand
ocean	opposite	season	set
slowly	songs	star	White

Traditional navigation

Without any modern equipment, the ancestors of Aboriginals, Torres Strait Islanders, Maoris and Polynesians travelled long **a**_____ across the oceans. They travelled as far west as Asia, north to the Hawaiian Islands, south to New Zealand and **b**_____ to Tahiti and beyond. Their navigation techniques were more an art than a science and learning the skills was a great **c**_____. They needed to know:

- how to read the weather

- about **d**_____ currents

- how wave patterns change near islands

- the different cloud types that form over islands, reefs and large land masses

- the behaviour of different birds, whales and **e**_____

- the paths of stars.

These people didn't have maps as we know them. Their maps showed mountains, large rocks, reefs and large trees but no **f**_____ directions. They lined up the key features with the path of stars they knew. Navigators memorised all the voyage details and used **g**_____ to help them remember. A good **h**_____ was a high-status, valued member of the community.

Alamy Stock Photo/Doug Houghton NZ

11. Applying trigonometry

Hawaiian navigators memorised a 'mind compass' that showed them where stars appeared to rise out of the ocean and sink back into the ocean as they set. The star compass includes information about the flights of **i**_____ and the direction of waves. This knowledge helped them find direction.

At its highest point, a zenith **j**_____ is directly overhead. Evidence suggests that early navigators knew the zenith star for islands they visited. The key navigational technique they used was known as 'back sighting'. Journeys began at the end of the day as the Sun began to **k**_____. The navigator lined up a prominent land feature **l**_____ him and followed the direction towards the destination's zenith star. It sounds simple but it isn't! Zenith stars can only tell you if you've reached the right **m**_____, not what course to follow once you've reached the right latitude.

Once the boat came to within 100 to 150 km of an island, the navigator could 'read **n**_____ and birds' to complete the final leg of the journey. Clouds over islands are higher, thicker, darker and move more **o**_____ than clouds over sea. Pinkish clouds form over reefs, greenish clouds form over forested islands and unusually bright clouds are reflecting sunlight off an atoll lagoon. Perhaps this is how **p**_____ got its name as 'The Land of the Long **q**_____ Cloud'.

In the morning, birds fly away from islands in search of food, returning at night. To locate an island, a navigator travels in the **r**_____ direction to the birds in the morning and in the same direction in the afternoon. Of course, the navigator has to be very aware of the bird's breeding **s**_____. When they have babies to feed, birds can fly to and from the island all day! Most likely, a bird carrying a fish is on its way back to its **t**_____.

2 Early navigators knew that the positions of the rising and setting Sun indicated the direction of east and west. Complete each statement.

 a If the setting Sun is behind you, you are sailing _____.

 b If you are sailing towards the setting Sun, you are travelling _____.

3 Early explorers used the Southern Cross for directions. The main axis of the Southern Cross points due south.

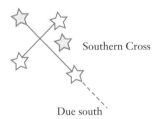

Southern Cross

Due south

Describe how the early explorers used the Southern Cross to find north.

4 Some Maori stories and songs provide a clue that early sea explorers may have followed migratory whales to travel between Samoa and New Zealand.

 a It's 2885 km from Samoa to Auckland. Migratory whales travel at an average speed of 7 km/h. Approximately how long did it take a canoe following migratory whales to travel from Samoa to New Zealand?

 b At the beginning of winter, whales pass Auckland on a bearing of $031°$ to swim to Samoa. What bearing do the whales follow to return to New Zealand from Samoa?

Trigonometry
crossword

TRIGONOMETRY FIND-A-WORD

There are 20 keywords from this chapter in the find-a-word puzzle below.

Copy the puzzle, look for keywords through the chapter, then see how many words you can find.

Write a list of the keywords, together with their meaning.

Q	W	E	C	O	S	R	T	D	Y	U	I
O	B	E	A	R	I	N	G	E	P	D	A
O	C	O	M	P	A	S	S	P	T	E	H
P	E	A	S	D	F	E	Z	R	A	G	Y
P	A	D	G	H	J	L	E	E	N	R	P
O	S	J	K	L	Z	E	N	S	X	E	O
S	T	A	C	S	V	V	I	S	S	E	T
I	N	C	L	I	N	A	T	I	O	N	E
T	M	E	M	N	Q	T	H	O	U	W	N
E	E	N	R	E	T	I	S	N	T	Y	U
U	I	T	R	U	E	O	T	O	H	P	S
M	I	N	U	T	E	N	A	N	G	L	E
A	N	O	R	T	H	S	R	W	E	S	T

SOLUTION TO THE CHAPTER PROBLEM

Problem

Early Australian builders knew that a large roof pitch helped prevent leaks and helped keep houses cool. The early roof pitches were often as large as 40°. Today, most roof pitches are about 22°.

How much higher was the roof of a 16 m wide early Australian building with a 40° roof pitch than a modern building with the same width and a 22° roof pitch?

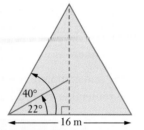

Solution

We can use trigonometry to calculate the height of each roof.

Early Australian roof

$$\frac{h}{8} = \tan 40°$$

$$h = 8 \times \tan 40°$$

$$h \approx 6.71 \text{ m}$$

Modern Australian roof

$$\frac{x}{8} = \tan 22°$$

$$x = 8 \times \tan 22°$$

$$x \approx 3.23 \text{ m}$$

Difference in heights $= 6.71 - 3.23$

$$= 3.48 \text{ m}$$

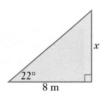

The roof of the early Australian building was 3.48 m higher than that of the modern building.

- Did any of the applications surprise you? If so, which ones?
- Which was the most interesting application?
- Do you know of any other trigonometry applications that weren't in this chapter? If yes, describe them to your teacher and friends.
- Are there any parts of this chapter you're not sure about? If yes, ask your teacher for help.

Copy and complete this mind map of the chapter. Use colours, diagrams and pictures to help you remember everything. When you've finished, ask your teacher to check your work.

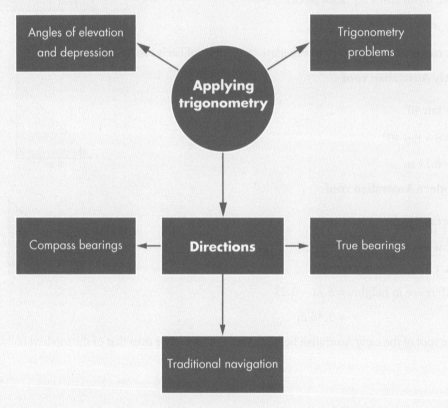

1 Find the value of each pronumeral, correct to one decimal place.

a

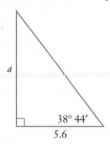

b

c

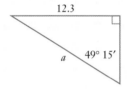

d

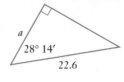

Exercise
11.01

2 Calculate the size of the angle marked θ, correct to the nearest minute.

a

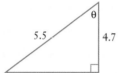

b

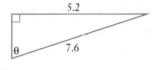

c

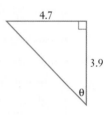

Exercise
11.01

3 The steps on the Great Wall of China are irregular. Some are tiny, whereas others are huge. In some places, the climb is quite challenging. In one section of the wall, the angle of inclination is 47° 15′ and the steps are narrow. Calculate the height, *h* cm, a person climbs up as they go across a horizontal distance of 60 cm. Answer to the nearest cm.

Exercise
11.02

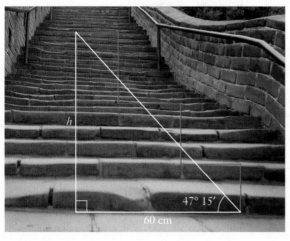

4 Australian building regulations for private dwellings state that the maximum rise in a step is 190 mm and the minimum tread width is 240 mm. Calculate the angle of inclination, to the nearest degree, of such a set of stairs.

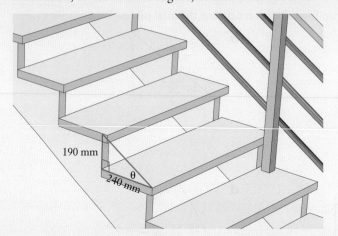

5 a State emergency workers are flying at a height of 500 m towards a mine disaster. The pilot can see the disaster site at an angle of depression of 11°. Calculate the horizontal distance from the helicopter to the disaster. Answer correct to the nearest 10 metres.

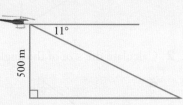

b Ginger the dog chased Oscar the cat up a tree. Ginger is 13 m from the base of the tree and her angle of elevation to Oscar is 24°. Ginger's eyes are 0.6 m above the ground. How high above the ground is Oscar? Answer in metres, correct to one decimal place.

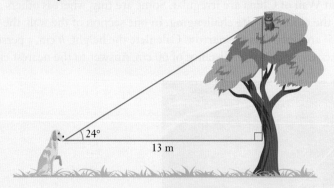

6 Which diagram shows the direction south-west?

Exercise
11.04

A

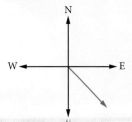

B

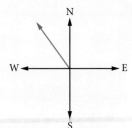

C

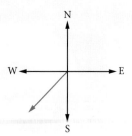

7 Match each bearing to its diagram **A**, **B** or **C**.

Exercise
11.05

   a   S 20° W   **b**   160°   **c**   340°

A

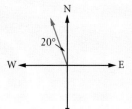

B

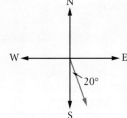

C

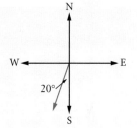

8 Jemma found a wild beehive 400 m east on a bearing of 136° from her home. Calculate the straight-line distance, *y*, from Jemma's home to the beehive. Answer to the nearest metre.

Exercise
11.06

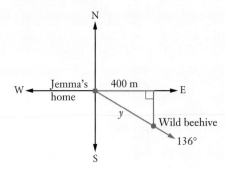

9 In the morning, a sea bird flew on a bearing of 236° away from an island towards a feeding ground. On what bearing will the bird fly to return to the island?

Exercise
11.07

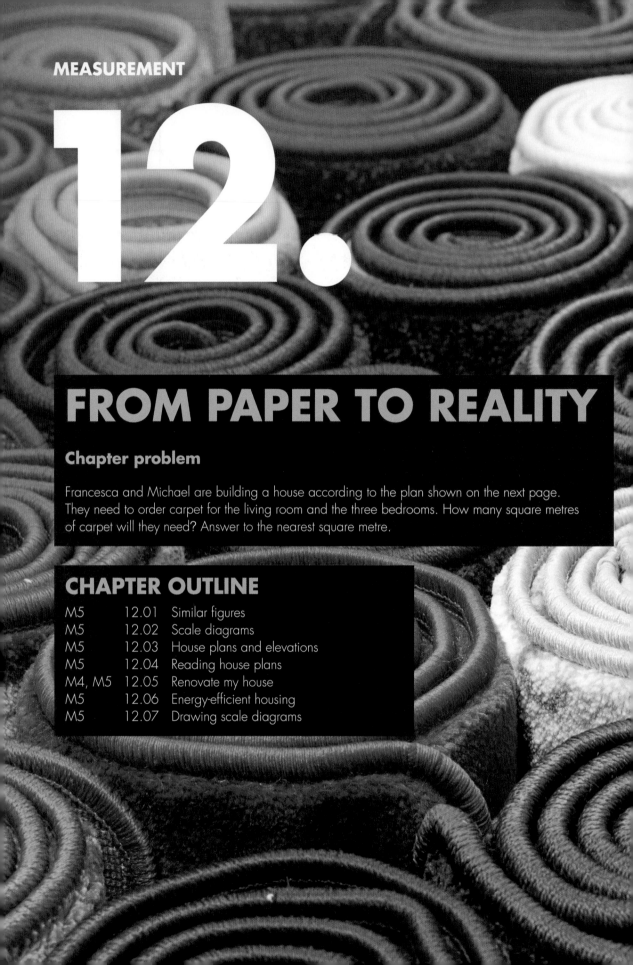

MEASUREMENT

12.

FROM PAPER TO REALITY

Chapter problem

Francesca and Michael are building a house according to the plan shown on the next page. They need to order carpet for the living room and the three bedrooms. How many square metres of carpet will they need? Answer to the nearest square metre.

CHAPTER OUTLINE

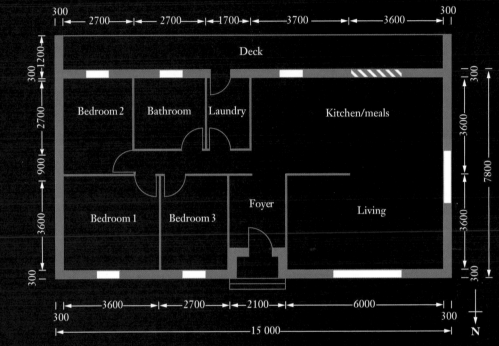

WHAT WILL WE DO IN THIS CHAPTER?

- Find unknown sides in similar figures
- Interpret and construct scale diagrams
- Find actual measurements from maps, house plans and elevations, such as lengths, perimeters and areas
- Calculate quantities and costs involving home renovations and energy-efficient housing
- Create scale drawings

HOW ARE WE EVER GOING TO USE THIS?

- Determining lengths on maps and house plans
- Building, renovating or buying a home
- Interpreting scale models and building plans
- Working in a building or painting trade

12.01 Similar figures

Applying
similar
triangles

Finding sides
in similar
triangles

Similar
triangles

Finding sides
in similar
triangles

These two photos have the same shape but are not the same size. One is an enlargement of the other, which is called an **image**. The photos are examples of **similar figures**.

◄──────── 6 cm ────────►

◄─────────────── 9 cm ───────────────►

iStock.com/ihsanyildizti

The **scale factor** of 2 similar figures is the amount that every length in the first figure is multiplied by to equal the corresponding length in the second figure.

EXAMPLE 1

a   What is the scale factor from the smaller photo to the larger photo?

b   Find the scale factor from the larger photo to the smaller photo.

c   What is the relationship between the length of a paddle in the larger photo to the corresponding paddle in the smaller photo?

Solution

a   The smaller photo is 6 cm long and the larger photo is 9 cm. Divide both sides of the equation by 6 cm.

$$6 \text{ cm} \times (\text{scale factor}) = 9 \text{ cm}$$
$$\text{Scale factor} = \frac{9 \text{ cm}}{6 \text{ cm}}$$
$$= 1\frac{1}{2}$$

The scale factor from the smaller photo to the larger photo is $1\frac{1}{2}$.

---

b   We reverse the order.
Divide both sides by 9 cm.

$9 \text{ cm} \times (\text{scale factor}) = 6 \text{ cm}$

$$\text{Scale factor} = \frac{6 \text{ cm}}{9 \text{ cm}}$$

$$= \frac{2}{3}$$

The scale factor from the larger photo to the smaller photo is $\frac{2}{3}$.

c   Every length in the larger photo is $1\frac{1}{2}$ times as long as the corresponding length in the smaller photo.

The length of the paddle in the larger photo is $1\frac{1}{2}$ times the length of the paddle in the smaller photo.

EXAMPLE 2

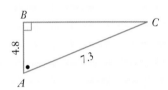 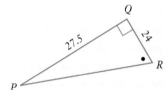

Triangles *ABC* and *RQP* are similar.

a   What is the scale factor from the smaller triangle to the larger triangle?

b   Calculate the length of *PR*.

c   How long is side *BC*?

Solution

a   Side *AB* in triangle *ABC* matches with side *QR* in triangle *PQR*. They each join the angles marked with a dot and a right angle.

$4.8 \times (\text{scale factor}) = 24$

$$\text{Scale factor} = \frac{24}{4.8}$$

$$= 5$$

b   Side *PR* matches with side *AC* because they are the hypotenuse of their triangles.

$AC \times (\text{scale factor}) = PR$

$7.3 \times 5 = PR$

$PR = 36.5$

c   Side *BC* matches with *QP*. They each join the right angle to the unmarked angle.

$BC \times (\text{scale factor}) = QP$

$BC \times 5 = 27.5$

$$BC = \frac{27.5}{5}$$

$$= 5.5$$

Exercise 12.01 Similar figures

1 For each pair of similar figures, calculate the scale factor from the left shape to the right shape.

a

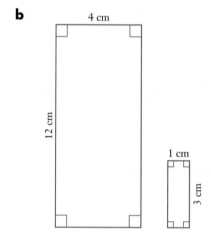

b

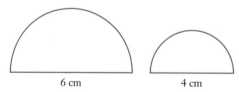

c

6 cm   4 cm

d

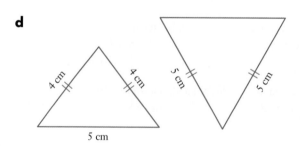

ISBN 9780170413534

2 In each pair of similar figures, find the scale factor from the left shape to the right shape and find the value of each pronumeral.

Example 2

a

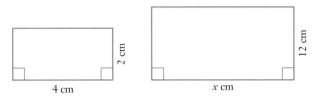

b

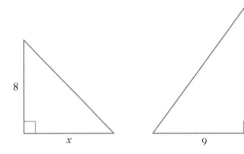

c

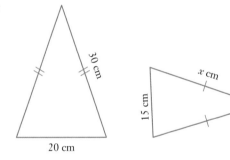

d

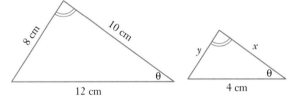

3 a What is the scale factor from triangle ABC to triangle DEC?

b Write a sentence to explain why side AC corresponds to side CD and calculate the length of CD.

c Find the length of side CE.

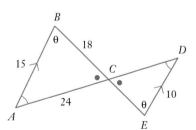

4 a Find the scale factor from triangle *BDC* to triangle *AEC*.

b Calculate the length of *EC*.

c How long is *ED*?

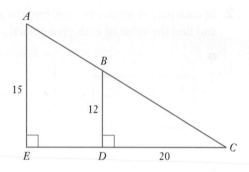

5 In the diagram, *AE* = *ED*.

a What is the scale factor from triangle *DEC* to triangle *DAB*?

b Calculate the lengths of *AB* and *CD*.

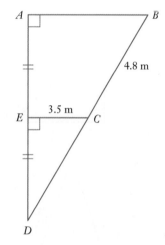

6 At the same time that the shadow of a 1.2 m high fence post was 1.5 m long, the shadow of a palm tree was 4.5 m long.

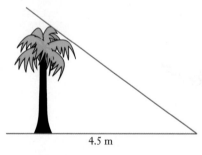
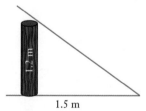

a What is the scale factor from the palm tree to the fence post?

b Calculate the height of the palm tree.

ISBN 9780170413534

CALCULATING HEIGHTS USING SIMILAR TRIANGLES

To complete this activity, you will need a 1 metre ruler, tape measure, a building, tree or tower with a shadow that can be measured, and a sunny day.

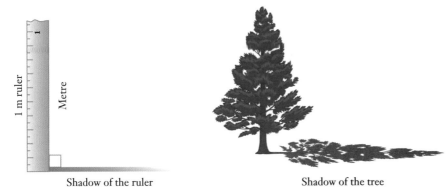

Shadow of the ruler   Shadow of the tree

1 Place the metre ruler vertically on the ground and measure the length of the ruler's shadow.

Vertical means at right angles with the ground.

2 Measure the shadow of the building, tree or tower.

3 Find the scale factor from the shadow of the metre ruler to the shadow of the building, tree or tower.

4 Calculate the height of the building, tree or tower.

12.02 Scale diagrams

Scaled vs
actual size

Scales and
scale
diagrams

Scale
drawings

Scale
diagrams

EXAMPLE 3

Barry, a farmer, used a scale of 1 cm : 100 m
when he drew this **scale drawing** of one of
his paddocks.

a   What are the actual dimensions of Barry's paddock?

b   Express the scale as a ratio.

Solution

a   The scaled length of the rectangle is   Actual length = 4 × 100 m
     measured to be 4 cm. Each centimetre   = 400 m
     represents 100 metres.

     The scaled width of the rectangle is   Actual width = 2.5 × 100 m
     2.5 cm.   = 250 m
                                                  Barry's paddock is 400 m by 250 m.

b   Change metres to centimetres, then   Scale = 1 cm : 100 m
     simplify.   = 1 cm : 100 × 100 cm
                                                        = 1 cm : 10 000 cm
                                                        = 1 : 10 000

EXAMPLE 4

Olga, an architect, made a scale diagram of a house using a scale of 1 : 50.

a   The actual length of the house is 20 m. What is the length of the house in the
     drawing?

b   The width of the house in the drawing is 24 cm. What is the actual width of the
     house?

Solution

a   Actual length = 20 m = 2000 cm

     Scaled length = 2000 cm ÷ 50 = 40 cm

b   Scaled length = 24 cm

     Actual length = 24 × 50 cm = 1200 cm = 12 m

Exercise 12.02 Scale diagrams

1 For each scale diagram, find the actual length shown by measurement and calculation, then express each scale as a ratio.

Example 3

a

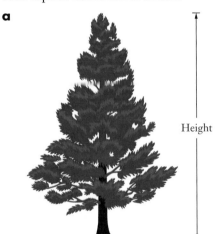

Height

Scale:
1 cm represents 2 m

b

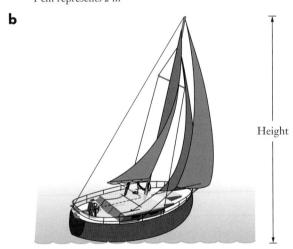

Height

Scale: 1 cm represents 5 m

c

← Width →

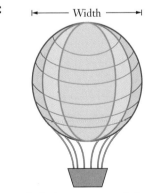

Scale: 1 cm represents 1.5 m

d

← Width →

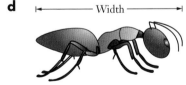

Scale: 1 cm represents 5 mm

e

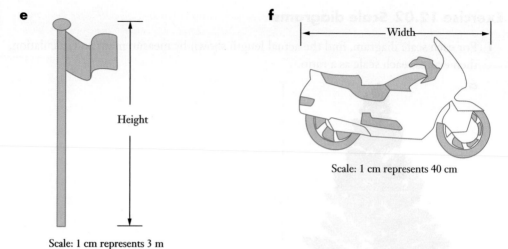

Height

Scale: 1 cm represents 3 m

Width

Scale: 1 cm represents 40 cm

2 A street map uses a scale such that 1 cm represents 200 m. Leanne walks from the Town Hall to the shopping centre, a distance that measures 5.5 cm on the map. How far does she walk?

3 Alan has a model of his favourite yacht in his study. The model has been made on a scale such that 2 cm represents 1 m. The model is 50 cm long. How long is the yacht?

4 Nagala used a scale of 1 : 50 when she constructed a scale drawing of her courtyard.

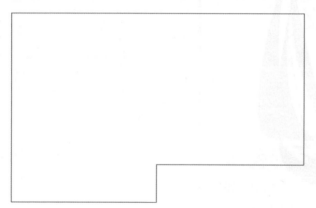

Find the length and width of Nagala's courtyard in metres.

5 Measure the length of each scaled image below, then use the scale ratio to calculate its actual length.

a Fish 1 : 3

Length

b House 1 : 300

Height

c Pen 1 : 4

Length

d Tennis racquet 1 : 16

Length

6 Lena collects model cars made in the ratio 1 : 43. Her favourite model is 9.3 cm long. Calculate the actual length of the car to the nearest metre.

7 A map has a scale of 1 : 50 000. Dean has to travel a distance measured as 128 mm on the map. How far does Dean have to travel? Express your answer in kilometres.

ISBN 9780170413534

12.03 House plans and elevations

A draftsperson drawing a design for a new building includes different views of the building.

- a **plan**: a diagram of the floor

- the **elevations**: views of the front, back, left side and right side of the building.

These diagrams show a garage with the front and right-side elevation and the view from above (plan):

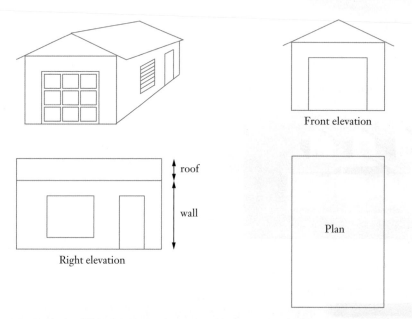

roof

wall

Right elevation

Front elevation

Plan

Shutterstock.com/Romakoma

Exercise 12.03 House plans and elevations

1 The plan and two elevations for Erik's workshed are shown below.

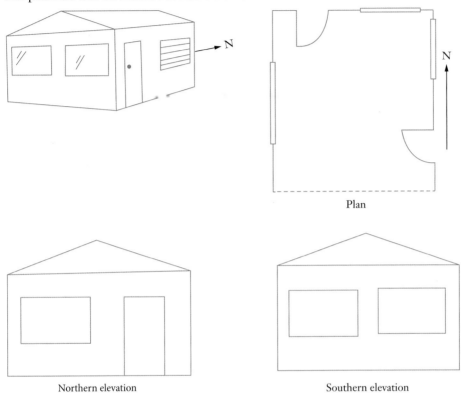

Plan

Northern elevation

Southern elevation

a The dotted line shows that the plan is incomplete. Copy and complete the plan.

b Draw the eastern and western elevations.

> Remember: the eastern elevation is the view from the eastern side.

2 For the house shown, sketch the front and side elevations. You do not need to draw the elevations to scale, but use a ruler and be as accurate as you can.

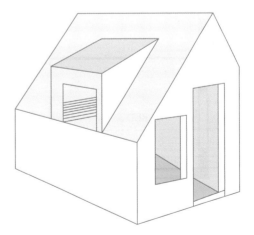

3 For the house shown, sketch the front and possible side elevations. You do not need to draw them to scale but use a ruler and be as accurate as you can.

Shutterstock.com/Ewelina Wachala

4 This is the plan of Grigor and Silvana's holiday house.

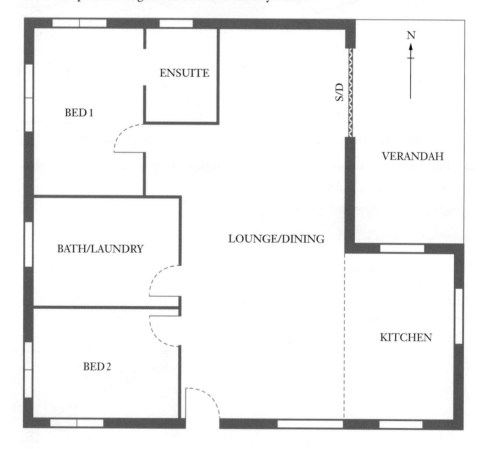

Sketch the northern and western elevations. You do not need to draw them to scale but use a ruler and be as accurate as you can.

5 This is the plan for a one-bedroom apartment, the front door of which faces north. Draw elevation views of each side of the house.

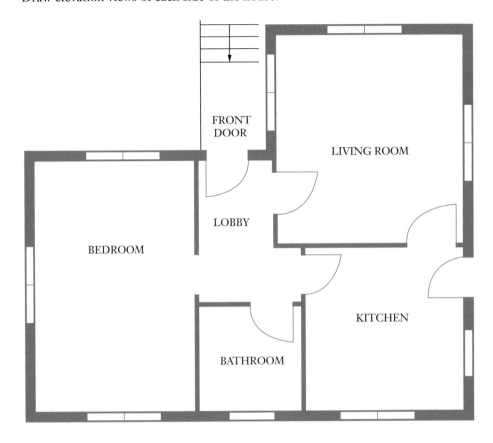

12.04 Reading house plans

House plans use a lot of different symbols and abbreviations. They are either drawn to scale or have measurements written on them. Often, measurements are shown in **millimetres** to avoid the use of decimal points, which can lead to errors in printing and reading.

EXAMPLE 5

This is the floor plan for Joe and Michelle's house.

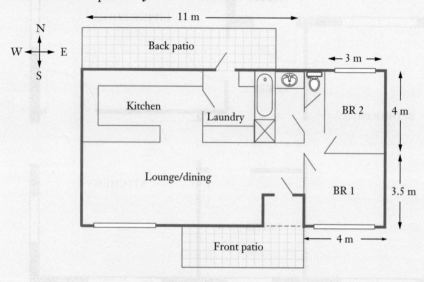

a   What are the dimensions of bedroom 1?

b   What is the area of bedroom 1?

c   What is the length across the front of the house?

d   What is the width of the house, excluding the patios?

Solution

a   Find bedroom 1 on the plan. Dimensions are the length and the width.

Dimensions of bedroom 1 are 4 m by 3.5 m.

b   Area = length × width.

$A = 4 \times 3.5$
$= 14 \text{ m}^2$

c   From the plan, find the full length.

Length = 11 + 4
$= 15 \text{ m}$

d   From the plan, find the full width.

Width = 4 + 3.5
$= 7.5 \text{ m}$

NCM 12. Mathematics Standard 1   ISBN 9780170413534

Exercise 12.04 Reading house plans

1 This is the plan for Menhal's new house. Measurements are shown in mm.

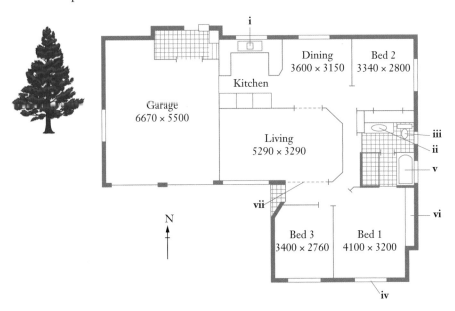

a What does each symbol used on the plan represent?

b Three areas on the plan are shaded with this pattern. What does this mean?

c How many bedrooms does the house have?

d Which bedrooms have built-in wardrobes?

e What are the dimensions of bedroom 1?

f Where is the laundry?

g Where is the linen press?

h How many bathrooms does the house have?

i How many toilets are there?

j How many doors lead into the bathroom?

k Which room has dimensions 5.29 m by 3.29 m?

l What are the dimensions of the garage?

m If Menhal looks out of the window of each room mentioned below, in which direction is she facing?

   i living room   **ii** dining room   **iii** bedroom 1

   iv bedroom 2   **v** garage

n Which direction do the garage doors face?

o If Menhal stands on the front door step, can she see into the bathroom?

p If Menhal stands inside the front doorway, list the four rooms she can see into.

q If Menhal is working in the kitchen, list the three rooms she can see into.

2 a How many bedrooms does this house have? Measurements are shown in metres.

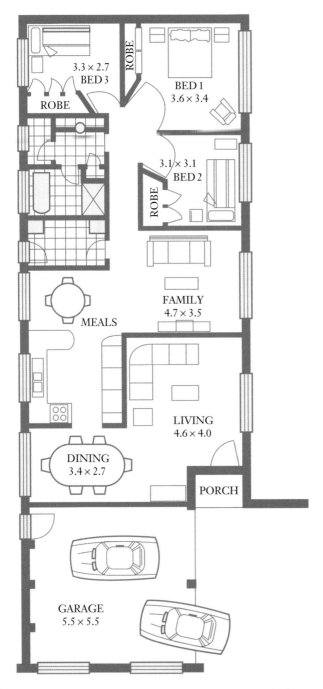

b How many of the bedrooms have built-in wardrobes?

c Is there an ensuite bathroom?

d The floor area of the house is 128 square metres. Calculate the cost of building this house at a rate of $672 per square metre.

3 This house plan has a scale of 1 : 100.

Measure and calculate the:

a length of the main bedroom

b length of the window in that room

c length of the laundry

d area of the bathroom

e longer side of the lounge room

f area of the dining room.

Scale 1 : 100

NCM 12. Mathematics Standard 1

ISBN 9780170413534

4 Judy and Keith are buying the house with this floor plan. Measurements are shown in mm.

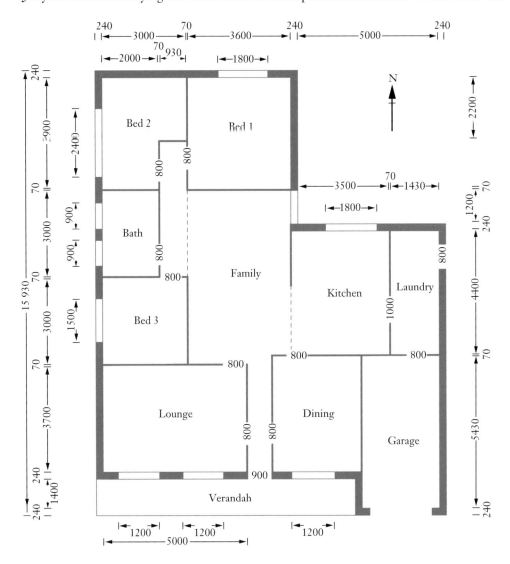

a What is the thickness of the external walls of this house in mm?

b What is the thickness of the internal walls?

c What is the front width of the house in metres?

d What is the length of the left side of the house?

e What are the dimensions of the family room?

5 a What are the dimensions of the family room on this plan? All measurements are in metres. Express your answer in mm.

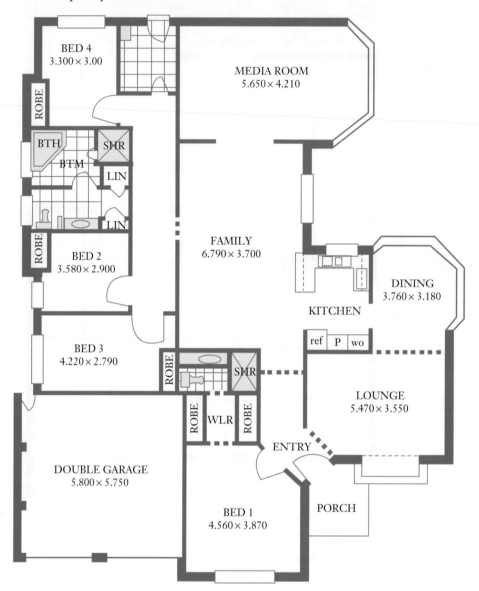

b How many toilets are there?

c Where are the 2 linen presses (cupboards for towels and sheets)?

d What does SHR stand for?

e Which bedroom has the smallest floor area?

f Which is bigger in area: the family room or the media room? By how many square metres?

iStock.com / pamspix

There are many websites that contain house floor plans. Search for 'project homes' on the Internet and choose some floor plans for houses that you like.

Google

For each of the plans:

• calculate the floor area of the whole house

• if possible, obtain an estimate for the cost of building the house

• calculate the cost per square metre.

Compare and contrast the features of each house.

Based on the information you have found above, decide which house you prefer and present your findings.

12.05 Renovate my house

We can use house plans and sketches to calculate the amount of materials required for a job and to estimate costs.

This is a diagram of Jackie's family room.

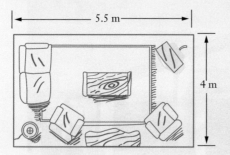

5.5 m

4 m

Jackie is laying cork tiles on the family room floor. The tiles cost $37.50 per square metre. How much will the cork tiles cost?

Solution

We need to find the area of the floor.

Area = 5.5×4
= 22 m^2

Multiply the area by the price.

Cost = $22 \times \$37.50$
= $825

Exercise 12.05 Renovate my house

1 The dining room in Nicole's home unit is very small. To make it look bigger, she plans to cover one wall with mirror tiles. The wall she plans to cover is 3 metres high and 2.5 metres wide.

a What is the area of the wall?

b The mirror tiles Nicole has chosen cost $36.50 per square metre. What will be the total cost of the mirror tiles for the wall in Nicole's dining room?

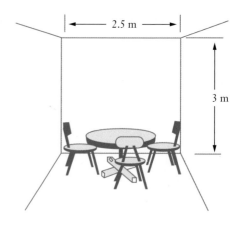

2.5 m

3 m

ISBN 9780170413534

2 Louis is replacing the skirting boards in his lounge room.

> Skirting boards are the strips of wooden boards that run around a room along the base of the walls.

a How many metres of skirting boards will he need?

b The skirting boards cost $11 per metre. Calculate the total cost of the skirting boards.

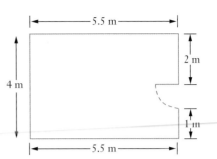

3 The floor area of Jesse's new house is shown.

a What is the floor area of this house?

b Using a cost of $840 per square metre, calculate the cost of building this house.

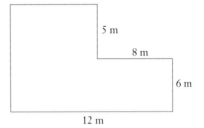

Use the floor plan for Joe and Michelle's house on page 362 to answer Questions **4** to **7**.

4 Joe and Michelle are renovating bedroom 1. The skirting boards need replacing.

a Calculate the perimeter of bedroom 1.

b The door is 1 m wide. How many metres of skirting board are needed?

c Skirting boards cost approximately $15 per metre. What will be the cost of the skirting boards?

5 Joe and Michelle are also replacing the carpet in bedroom 1.

a How many square metres of carpet will they need?

b Carpet comes in rolls 3.6 m wide. What length of carpet will Joe and Michelle need?

c The carpet they like costs $79.45 a metre. How much will the carpet cost in total?

6 Joe and Michelle also need to paint the ceiling.

a What is the area of the ceiling?

b The ceiling requires two coats of paint. One litre of paint covers approximately 12 m^2 of the ceiling.

   i Calculate the number of litres of paint they require.

   ii The ceiling paint costs $34.90 for a 4 litre can. Estimate the cost of paint for the ceiling.

7 a In which order should Joe and Michelle do the renovations from Questions **4**, **5** and **6**? Give reasons for your answer.

b What is the approximate total cost of these renovations?

Use the floor plan for Judy and Keith's house from page 367 to answer Questions **8** to **11**.

8 Judy and Keith have decided to replace all the floor coverings before they move into the house. They are going to lay wood parquet flooring in the lounge room. The flooring costs $68 per square metre. How much will the flooring cost?

9 Other rooms need new carpet. Judy has chosen a carpet that is 3.66 m wide. How many metres of carpet are required for each of these rooms?

   a bedroom 1

   b bedroom 2

   c bedroom 3

10 The kitchen and laundry are to have new tiled floors. The tiles cost $41.10 per square metre, which includes an allowance for breakage and wastage. How much will it cost to buy tiles for the floor? Answer correct to the nearest dollar.

11 Judy and Keith are expecting a baby and they are going to make bedroom 2 the baby's room. Judy wants to decorate the room with a frieze pattern around the walls. It will go above the door and the windows.

   a How many metres long will the frieze be?

   b Each roll of frieze is 5 m long. How many rolls of frieze will Judy need to buy?

iStock.com/ac_bnphotos

12 Judy and Keith were surprised by the estimated cost of their renovations. Before signing off on the purchase of this house, they checked the cost of building a project home of a similar style and with the same floor area as this house. The cost of the project home was approximately $594 per square metre. Calculate the estimated cost of building this project home.

12.06 Energy-efficient housing

When we build a house, these are some of the important things to consider:

- Location of the house on the site: the long axis of the house should run east-west

- Internal design – living areas should be on the north side to capture winter sun.

- Insulation – this is the most effective way to reduce energy use for heating and cooling.

- Windows/skylights – place them to maximise winter sun

The following diagrams illustrate Kim and Chris' house.

The **site plan** below shows the block of land and the house on it. The solid lines are the outline of the house and the dotted lines are the outline of the roof. All measurements on these plans are in millimetres.

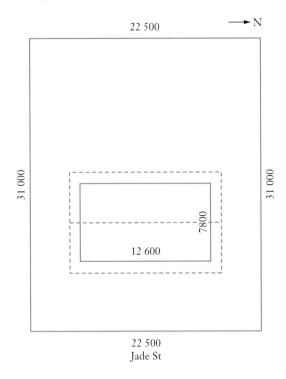

The floor plan shows the house design.

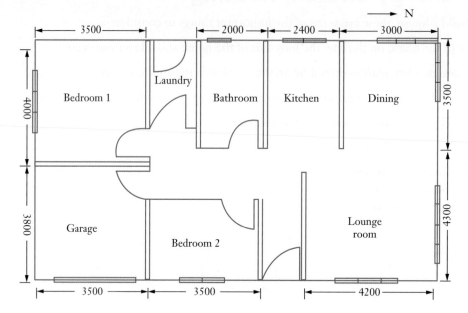

The northern elevation shows one side of the house.

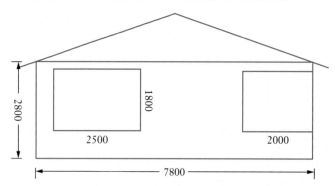

Exercise 12.06 Energy-efficient housing

1 a Is Kim and Chris' house sited in the most energy-efficient way?
   b Are the living areas facing north?

2 Using the site plan:
   a find the area of the block of land in m^2
   b find the floor area of the house in m^2.
   c hence, calculate the area left for gardens and lawns.

3 The eaves extend 1 m beyond the house.
   a Draw a neat sketch of the plan view of the roof area, showing all measurements.
   b Calculate the roof area in m^2.

4 When insulation is installed in a house, it usually covers the ceiling of the house, not the eaves or the garage.

   a What is the ceiling area of Kim and Chris' house in m^2?

   b A bag of ceiling insulation covers 9 m^2. How many bags of insulation do Kim and Chris need to buy to put ceiling insulation in the house?

   c Each bag costs $52.60. Calculate the cost of insulating the ceiling.

5 Kim and Chris are planning to purchase reverse-cycle air conditioning for the lounge and dining area.

   a Using the floor plan, calculate the floor area of these two rooms in m^2.

   b What percentage of the house floor area is this?

   c Given the information in the table below, what size reverse-cycle air conditioner would you recommend Kim and Chris buy for this area?

Floor area	Size of air conditioner
Up to 17 m^2	2.6 kW
Up to 25.5 m^2	3.9 kW
Up to 34 m^2	5.2 kW
Up to 42.5 m^2	6.5 kW
Up to 51 m^2	7.8 kW

   d The above table for air conditioners assumes the usual ceiling height of 2400 mm in the rooms to be air conditioned. Give an example of a situation where you could *not* use this table to choose an air conditioner.

6 Often, skylights are installed on the southern side of a house to improve the light in those rooms. On this plan, which room do you think could benefit from a skylight?

7 On the northern elevation:

   a calculate the area of the two windows in m^2

   b calculate the area of the northern wall in m^2, including the windows.

   c what percentage of the wall is window?

8 According to Australian building regulations, the window area of any house should be 25% or less of the floor area. Give an estimate for the maximum window area possible for this house.

9 The owners decide to plant 35% of the land not covered by the house with native plants to reduce the water requirements of the outside area.

   a Using your answer to Question **2** part **c**, how many square metres is 35% of the land not covered by the house? Answer to the nearest square metre.

   b Draw a neat sketch of the site and show a possible placement for the native garden.

12.07 Drawing scale diagrams

To make a scale diagram, start with a rough sketch with the required measurements. Then, choose a scale and accurately draw the diagram.

> To calculate a scaled length, divide the actual length by the scale.

EXAMPLE 7

This is a sketch of Farmer Freda's field. Draw a scale diagram of the field.

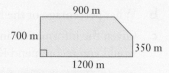

Solution

Choose a suitable scale – we want it to fit easily on the page.

Use the scale 1 cm represents 200 m.

Calculate the scaled length for each measurement by dividing by the scale.

$900 \div 200 = 4.5$ cm
$700 \div 200 = 3.5$ cm
$1200 \div 200 = 6$ cm
$350 \div 200 = 1.75$ cm

Use these measurements to draw your scale diagram.

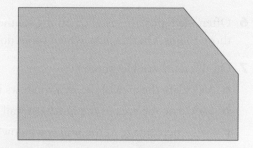

EXAMPLE 8

Sue is going on a bushwalk from her camping site. She walks 2.5 km due east and then 1.9 km northwest.

a Draw a scale diagram of Sue's walk.

b Measure the diagram and calculate how far Sue is from her campsite.

We can use our scale drawing to find other measurements.

Solution

a Choose a suitable scale.

Use scale 1 cm represents 0.5 km.

Find the scaled measurements by dividing by the scale.

$2.5 \div 0.5 = 5$ cm
$1.9 \div 0.5 = 3.8$ cm

Use a ruler and a protractor to construct the scale diagram.

Northwest is 45° (halfway) between west and north.

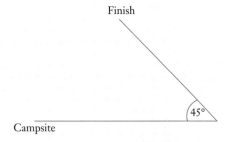

b Join the starting point to the finishing point and measure this line.

The line is 3.5 cm long.

Find the actual distance by multiplying by the scale.

$3.5 \times 0.5 = 1.75$ km
Sue is 1.75 km from her campsite.

Exercise 12.07 Drawing scale diagrams

You will need a ruler and a protractor to complete this exercise.

1 Make a scale drawing of this field.

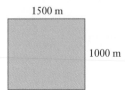

2 Andrew drew a sketch of a courtyard. Make a scale drawing of the courtyard.

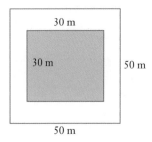

3 Samantha designed this cutting plate. Make a scale drawing of her cutting plate.

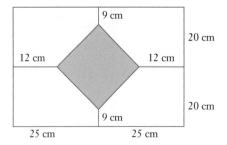

4 Hamish and Jacob are setting off on a hike. They walk 5 km due west of their starting point and then turn to walk 7 km south. They stop for lunch and then walk another 6 km in a northeasterly direction before stopping for afternoon tea.

   a Draw a scale diagram of Hamish and Jacob's walk.

   b How far are they from their starting point?

5 **a** Construct a scale drawing to calculate the actual height of the hot air balloon above the ground.

   b Use trigonometry to calculate the height of the balloon, correct to one decimal place.

   c How accurate was your scale drawing?

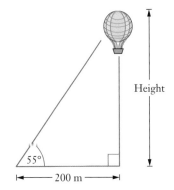

Height

55°

200 m

6 An old wooden gate 80 cm wide by 120 cm high needs a diagonal brace for support.

   a Construct a scale drawing of the gate and find the actual length of the brace.

   b Use Pythagoras' theorem to check the answer you obtained from the scale drawing. Answer correct to one decimal place.

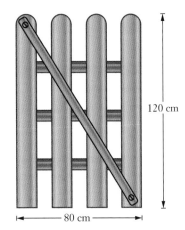

120 cm

80 cm

7 Jodie wants to swim across the river. Draw this diagram to scale and calculate the width of the river.

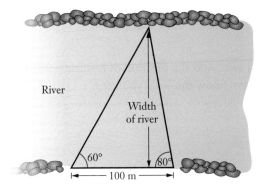

River

Width of river

60° 80°

100 m

8 A golf course has a large lake as an obstacle. Most golfers follow the dog leg around the lake. How far is it straight across the lake, from the tee to the hole? Use a scale diagram to answer this question.

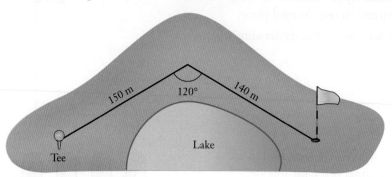

INVESTIGATION

MAKE MY OWN SCALE DRAWING

Part A

1 Choose something that is a large rectangle; for example, a paved area, a quadrangle, a brick wall, a whiteboard or the classroom floor.

2 Measure the length and width of the rectangle.

3 Make a scale drawing of your rectangle. Be sure to include the scale you used.

Part B

1 Measure the length and width of the school reception area or foyer.

2 There are plans to enlarge the length and width of this area by 50%. What will be the new dimensions of this area?

3 Draw a scale diagram of the enlarged area. Add the position of furniture and any other items in the school reception area. Show the scale you used.

WORD MATCH

Match each word in the left column to its correct meaning in the right column.

	Word		Meaning
1	Dimensions	A	For an enlargement or reduction, the image length divided by the object length
2	Eaves	B	A ceiling window that allows additional light into a room
3	Elevation	C	The relationship between a real object and its scale diagram
4	Energy-efficient	D	A drawing showing where a building is on a block of land
5	Enlarge	E	A unit of length used in house plans
6	House plan	F	To make a drawing or object larger
7	Insulation	G	Figures with the same shape but not the same size
8	Milllimetre	H	Icons that illustrate features on a house plan
9	Reduce	I	The length and width of a room on a plan
10	Scale	J	A diagram showing the arrangement and measurement of rooms in a house
11	Scale drawing	K	Material placed in the ceiling to help keep a house cool or warm
12	Scale factor	L	The view of a building from the side
13	Similar figures	M	Using as little energy as possible
14	Site plan	N	The space between the house and the edges of the roof
15	Skylight	O	Represents real objects that are too big or too small to draw with the actual measurements
16	Symbols	P	To make a drawing or object smaller

SOLUTION TO THE CHAPTER PROBLEM

Problem

Francesca and Michael are building a house according to this plan. They need to order carpet for the living room and the three bedrooms. How many square metres of carpet will they need? Answer to the nearest square metre.

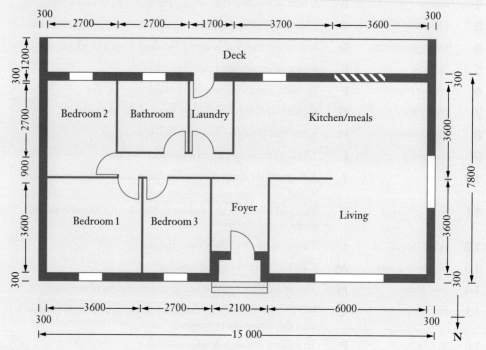

Solution

Change the measurements for each room into metres and calculate its area.

Area of living room = 6000 mm × 3600 mm
$$= 6 \text{ m} \times 3.6 \text{ m}$$
$$= 21.6 \text{ m}^2$$

Area of bedroom 1 = 3600 mm × 3600 mm
$$= 3.6 \text{ m} \times 3.6 \text{ m}$$
$$= 12.96 \text{ m}^2$$

Area of bedroom 2 = 2700 mm × (2700 mm + 900 mm)
$$= 2.7 \text{ m} \times 3.6 \text{ m}$$
$$= 9.72 \text{ m}^2$$

Area of bedroom 3 = 2700 mm × 3600 mm
$$= 2.7 \text{ m} \times 3.6 \text{ m}$$
$$= 9.72 \text{ m}^2$$

Total area = 21.6 + 12.96 + 9.72 + 9.72
$$= 54 \text{ m}^2$$

ISBN 9780170413534

12. CHAPTER SUMMARY

- What parts of this chapter were new to you?
- Give examples of jobs where you would use the skills learned in this chapter to draw scale diagrams or read house plans.
- Write any difficulties you had with the work in this chapter. Ask your teacher to help you.

Copy and complete this mind map of the chapter. Use colours, diagrams and pictures to help you remember everything. When you've finished, ask your teacher to check your work.

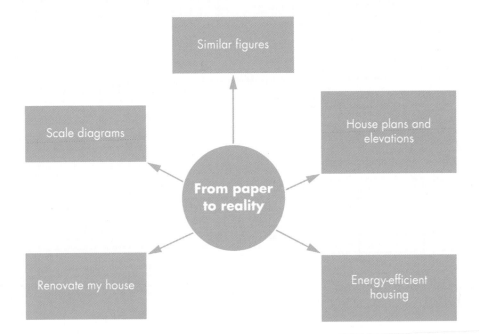

12. TEST YOURSELF

Exercise
12.01

1 For each pair of similar figures:

    i find the scale factor from the left figure to the right figure

    ii find the value of each pronumeral.

a
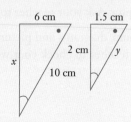

b
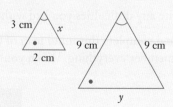

c
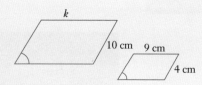

d
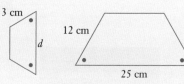

e
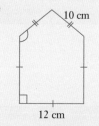

Exercise
12.02

2 Measure the length of each scale drawing below, and then use the ratio to work out the actual length of the object shown.

    a Fish

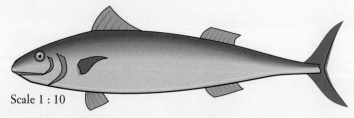

Scale 1 : 10

    b Frog

Scale 1 : 4

3 a Draw elevation views of the northern and western sides of the house shown below.

You do not need to draw them to scale, but use a ruler and be as accurate as possible.

Exercise 12.03

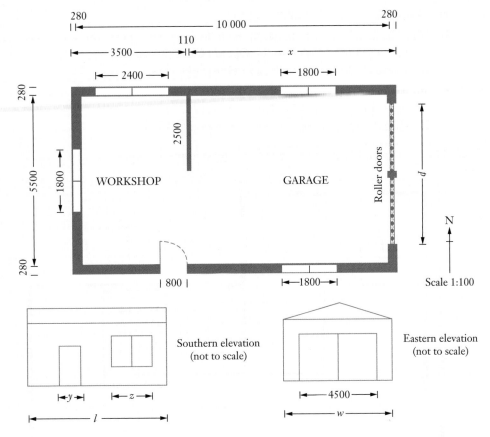

Southern elevation
(not to scale)

Eastern elevation
(not to scale)

Scale 1:100

b What is the length of x on the plan?

c Find the lengths of y, z, l, w and d.

4 Use the plan in Question **3** above to answer the following questions.

Exercise 12.04

a What is the width of the external walls of the building?

b What is the width of the wall between the workshop and the garage?

c Find the internal width of the building.

d What are the dimensions of the workshop?

e What are the dimensions of the garage?

5 On a tourist map of Sydney, the scale is given by:

0   500 m

a Write this scale as a simplified ratio.

b Find the actual distance between the following places given the scaled distance.
  i Circular Quay station to the Opera House (2.5 cm)
  ii Pyrmont Bridge to Parliament House (4.4 cm)

c Find the scaled distance between the following places given the actual distance.
  i Art Gallery of NSW to Sydney Tower (875 m)
  ii Circular Quay station to Central station (2.5 km)

6 This is the plan for the house that Sam and Nic are buying.

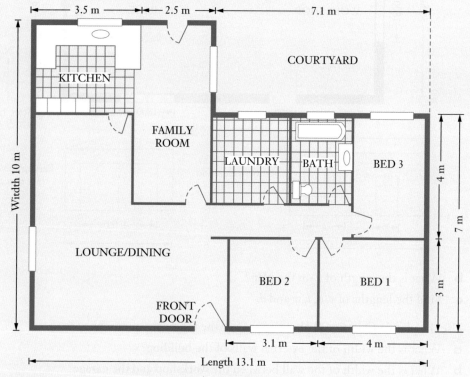

a How many doors are shown on this plan?

b What are the length and width of the house?

c Calculate the width of the kitchen.

d **i** Find the floor area of the house.

  ii An estimate for the cost to build a house is $672/m². Estimate the cost of building this house.

e What percentage of the floor area of the house is taken up by bedroom 1? Answer correct to one decimal place.

f A builder is going to tile the courtyard. What is the area of the courtyard?

g To allow for cutting and fitting, a builder always buys 10% more tiles than required to cover a floor area.

   i How many square metres of tiles should the builder buy for the courtyard, including the extra 10%?

   ii If each tile measures 32 cm by 32 cm, how many tiles are needed in total?

h The guttering across the front of the house needs replacing.

   i How much guttering is needed?

   ii The guttering company charges a fee of $50 plus $24.75 per metre to supply and install the guttering. How much will it cost to gutter the front of the house?

7 This is the floor plan for Susan and Ian's house. They have decided to renovate some areas of the house.

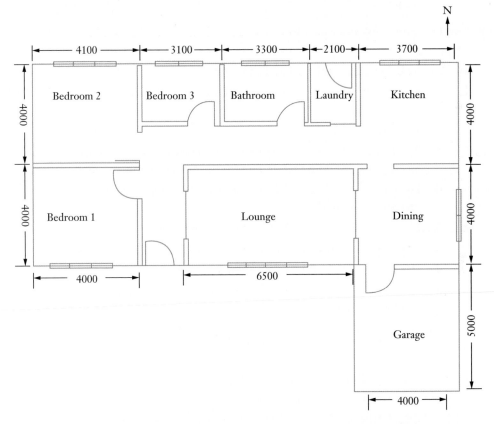

a Ian and Susan are going to lay wood parquet flooring in the lounge room. The flooring costs $68 per square metre. How much will the flooring cost?

b They are going to replace the skirting boards in bedroom 1. Assume the door is 820 mm wide.

   i How many metres of skirting board are needed?

   ii Skirting boards cost approximately $15 per metre. What will be the cost of the skirting boards?

c The tiles on the floor of the bathroom and laundry need replacing. The width of these rooms is 2300 mm.

   i Calculate the total floor area of the bathroom and laundry.

   ii Each tile is 200 mm by 200 mm. How many tiles are needed?

   iii The tiles cost $41.10 per square metre allowing for wastage. How much will the new tiles cost?

d The ceiling of the lounge room needs to be repainted.

   i What is the area of the ceiling?

   ii The ceiling requires two coats of paint. One litre of paint covers approximately 12 m^2. How many litres of paint will be required to paint the ceiling?

   iii Ceiling paint costs $34.90 for a 4 litre can. Estimate the cost of painting the ceiling.

8 Use Ian and Susan's floor plan from the previous page and the list of energy-efficient features given on page 373 to answer these questions.

a Are the living areas of the house facing north?

b Ian and Susan wish to install insulation in the ceiling of the house. This *does not* include the garage.

   i What is the ceiling area of this house in m^2?

   ii A bag of ceiling insulation covers 9 m^2. How many bags of insulation are needed for this house? Round up to the nearest whole number.

   iii Each bag costs $52.60. Calculate how much Susan and Ian would pay to install the insulation in the ceiling.

c There is air conditioning in the lounge and dining area.

   i Using the floor plan, calculate the floor area of these two rooms in m^2.

   ii What percentage of the house floor area is this?

   iii Given the information in the table, what size air conditioner should Susan and Ian buy for this area?

Floor area	Size of air conditioner
Up to 17 m^2	2.6 kW
Up to 25.5 m^2	3.9 kW
Up to 34 m^2	5.2 kW
Up to 42.5 m^2	6.5 kW
Up to 51 m^2	7.8 kW

 ISBN 9780170413534

9 Neil and Ted are keen scuba divers. When they moor their boat, the angle between the anchor rope and the top of the water is 30°.

a Construct a scale drawing to calculate the required length of anchor rope if the water is 20 m deep.

b Use trigonometry to check your answer to part **a**.

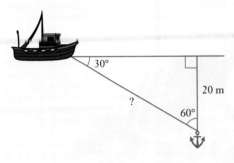

13.

UNBIASED DATA

Chapter problem

Avril is employed by a large car manufacturer to research the available optional extras that customers want in a new car. How is she going to do this?

CHAPTER OUTLINE

WHAT WILL WE DO IN THIS CHAPTER?

- Understand the process of statistical investigation
- Design a questionnaire correctly and consider target populations
- Detect bias in questionnaires and samples
- Review methods of analysing data and drawing graphs
- Compare two sets of data
- Communicate the findings of a statistical investigation and make recommendations

HOW ARE WE EVER GOING TO USE THIS?

- When we read or consider information in the media
- When we evaluate the results of surveys
- When we undertake a survey

13.01 The statistical investigation process

Information is collected by a variety of groups to enable people to make informed decisions. This process is called **statistical investigation** and involves the following steps.

- **Identifying a problem and posing questions** – deciding what information is needed and in what form.

- **Collecting data** – identifying the target population and choosing to use a **census** or a **sample** of the population. We must decide how we will collect information and then collect the information.

- **Organising data** – after we collect the data we need to organise it. We could use a frequency table, with either grouped or ungrouped data, or a spreadsheet.

- **Representing data** – in this step, we present the information in a way that makes it easy to follow and understand. Often, we do this graphically.

- **Analysing data** – this step involves the calculation of summary statistics, such as mean, median and mode, then analysing the assembled information.

- **Interpreting and communicating the findings** – this step involves presenting the data in an understandable way with the conclusions supported by the statistics.

Exercise 13.01 The statistical investigation process

1 In your own words, summarise the steps in the process of statistical investigation.

2 Classify each activity below as one of these steps:
- posing questions (PQ)
- collecting data (CD)
- organising data (OD)
- representing data (RD)
- analysing data (AD)
- interpreting/communicating findings (ICF)

a Liong asks fellow students in Year 12 how they travel to school

b Jane draws a frequency histogram of her data

> **Frequency histogram** is a bar chart (column graph) that shows the frequency of a single score or group of scores.

c Anna finds the mean house price for her suburb

d Managing Director Theo decides he needs to know customers' favourite car colour

ISBN 9780170413534

e Kieran makes recommendations in his final report

f Emilia puts her information on favourite holiday destinations into a frequency table

g Marian concludes that coffee is her friends' favourite daytime drink

h Simon visits the houses in Lawson Street to collect their Census forms

i Madeline decides to research the drinking habits of 18-year-olds

j Liam displays data in a sector graph about pocket money received by Year 11 students

k Phoebe writes a report on her data about popular sports

l Kim goes through a pile of surveys and records the responses to a specific question

3 Choose a topic to research and describe how you would implement each of the 6 steps of a statistical investigation.

4 Some organisations collect information on a large scale. Visit the website of each organisation below and write at least five things about which they collect information.

Australian Bureau of Statistics

    a Australian Bureau of Statistics (ABS)

    b United Nations (UN)

United Nations

    c World Health Organization (WHO)

5 The privacy of collected information can be important. Go to the ABS website and search for the **Census** page. Scroll down and click on **Privacy, Confidentiality & Security**. Describe in your own words how the ABS ensures the privacy of the information we provide in the Census.

World Health Organization

13.02 Questionnaires

Once we have identified the problem or topic we want to investigate, using questionnaires is one of the most common ways to collect information. A good questionnaire has the following features.

- It uses simple language.

- Questions are written so that the meaning is clear.

- There is only one part to each question so it is easy to answer.

- It meets privacy requirements. Unauthorised people cannot access your information.

- It is free from bias.

The style of answers and any provided alternatives are another important consideration.

Exercise 13.02 Questionnaires

Andrew is a market researcher. The United Club has asked him to evaluate the facilities and services it provides for members. He has designed the first draft of a questionnaire to survey club members about the activities the United Club provides. Andrew's draft is given below. Read the questionnaire and answer the following questions.

United Club   **Help us help you!**

Please answer this short questionnaire to help the United Club better serve you.

1 How old are you?   **2** Are you male or female?
☐ Less than 30 years   ☐ Male
☐ Between 31 and 40 years   ☐ Female
☐ Between 41 and 50 years
☐ Over 50 years

3 Which of the Club's entertainments do you attend?
_____

4 How often do you attend the following Club activities?
Concerts   Films   Dance nights   Trips away

5 Do you think the entertainment offered is reasonably priced?

6 Should the club offer more types of entertainment and, if so, what should it offer?

7 Are there any other comments you would like to make about the Club's entertainment program?

Thank you for taking the time to complete this questionnaire. Please hand it in at the bar.

---

1 **Open-ended questions** require the person answering to write their own answer. How many questions in the questionnaire are open-ended?

2 **Closed questions** give a choice of options for the answer. How many questions in the questionnaire are closed?

3 The first question about age has a problem with the options given. What error has Andrew made? (Hint: where would a person aged 30 tick?). Rewrite the options to this question to fix this error.

4 When Andrew asks 'Which of the Club's entertainments do you attend?' it is not clear to which of the Club's activities Andrew is referring. Rewrite the question to make it clear.

5 When Andrew asks 'How often do you attend the following Club activities?' he will get a range of answers. Analysing the data will be easier if Andrew provides some options. Suggest a series of tick boxes that Andrew could provide for people to record their answers.

6 Would all Club members agree on what 'reasonably priced' means? Suggest how Andrew could change this question to make it clearly understood.

7 Write another question about the cost of entertainment for this questionnaire.

8 There is a risk that answers won't remain private if the questionnaires are handed in at the bar. Suggest a more secure way to collect the questionnaire forms.

9 Do you think Andrew has missed anything in his questionnaire? (Hint: How would people who don't attend entertainment express their views?) Write 3 additional questions Andrew could use.

10 Using questionnaires is not the only way to collect data. In what other ways can information be collected?

13.03 Sample vs census

Once we have identified the problem or topic we want to investigate and have planned our questionnaire, we must decide the group of people we need to ask to get the information we need. This is called the **target population**.

In Year 11, we learned about the difference between taking a census and using a sample to collect information. We need to decide which option is best for the question we are asking.

EXAMPLE 1

The Year 12 Formal organising committee wants to know the preferred food to serve.

a What is the target population for this topic?

b Is a census or a sample the better option for this investigation?

Solution

a	The target population is all those who will be attending the formal.	The target population is Year 12 students, teachers and parents.
b	This would depend on the number of students in Year 12.	Small Year 12: Use a census Large Year 12: Use a sample

Exercise 13.03 Sample vs census

1 For each investigation, who is the target population?

    a   Favourite movies of teenagers in NSW

    b   Hours of paid work per week completed by students in your school

    c   Number of children in each family in your street

    d   Zengage On-Demand TV wants to know how satisfied its customers are with its service

    e   NCM Bank wants to know if its Internet banking customers find it easy to use

    f   The venue Year 12 wants to use for their formal

    g   Most popular TV shows

    h   Time taken to deal with complaints to a major communications company

    i   In NSW schools, boys complete more homework than girls

    j   Council wants to know what recreational equipment should be added to the town's parks

    k   Voting intentions at the next Federal election

    l   Number of people using emergency departments in hospitals across NSW on Saturday nights

2 For each investigation in Question **1**, is a census or a sample the better option?

13.04 Bias

Bias is an unwanted influence in sampling or questionnaires that makes the results unreliable.

In Year 11, we looked at possible bias in sampling that unfairly favours a particular section of the **population** and produces data that does not accurately reflect the whole population.

The way a question is worded in a questionnaire can introduce bias. The wording of the questions can lead people towards a particular view. This can unfairly influence the way people answer the questions.

> ### EXAMPLE 2
>
> Andrew, our market researcher from Exercise 13.02, has written some questions to ask Club members about the Club bistro.
>
> **a**   How often do you eat at our fabulous bistro?
>
> **b**   Rate your last meal: Great   Yummy   Quite nice
>
> In what way are these questions biased? Rewrite them so they are not biased.

Solution

a Using the word 'fabulous' encourages the person answering the question to give a positive answer.
It would be better to ask: 'How often do you eat at the bistro?'

b This question doesn't give any negative options. An unbiased question could be:
'Rate your last meal:
Poor   Average   Good'

Exercise 13.04 Bias

1 For each question below:

Example 2

   i   How is it biased?

   ii Rewrite it so it is not biased.

a Do you prefer to holiday in fascinating Sydney or in wet Melbourne?

b Do you agree that we should increase the tax on those disgusting cigarettes?

c Rate the last movie you saw: OK   Good   Fantastic

d Do you prefer exciting rugby league or boring old rugby union?

e Do you think the Club is open long enough or should we be able to rage on into the early morning?

f Rate your health fund: useless   not much good   reasonable

g Are there enough events for young people in this boring town?

h Isn't this just the greatest movie you've ever seen?

i Are you one of those people who walk to school?

j Rate the gym you attend: best ever   good   OK

2 As part of their assessment, Ms Benedict's Year 12 class is going to undertake a number of investigations. The students need to decide who they will ask to answer their questions. For each investigation suggest:

   i   What sort of group should be surveyed to get useful results?

   ii What sort of group is likely to give less useful or biased results?

Possible investigations:

a New tourist attractions that could be built in Coffs Harbour

b Children's favourite holiday destinations

c Most popular teacher at Nelson Lakes High

d Most popular sport in Australia

e Changes to the school uniform

f Exercise equipment to be installed in outdoor areas

g Shops to include in a new shopping centre

h Entertainment provided in a chain of retirement homes

13.05 Organising and displaying data

We have completed the first steps of the statistical investigation process: chosen a question, constructed a well-worded questionnaire and selected an unbiased sample to complete our questionnaire.

The next step is to collect the data: get our chosen sample to answer the questionnaire.

Once we have collected the data, we need to organise it and present it in a way that is easily understandable.

In Year 11, we looked at putting data into frequency tables and drawing appropriate graphs. Ask your teacher for some hints if you can't remember what to do in any of the following questions.

Exercise 13.05 Organising and displaying data

1 At New Century Motors, customers were asked: 'What is your preferred car colour?' Their answers were coded as follows:

black = B, silver = S, white = W, red = R, other colours = O.

This is the data that was collected during one week.

B	B	W	R	W	R	B	S	O	W	R
B	S	B	O	W	B	W	S	B	B	R
R	O	W	B	B	O	O	R	B		

a Copy and complete this frequency table for the data.

Colour	Tally	Frequency
Black		
Silver		
White		
Red		
Other		

b Display this data as a column graph.

c What is the most popular colour among New Century Motors' customers?

2 Hadieya counted the number of phone calls she made each day for the month of June. The data is shown below.

5	3	6	2	4	3	3	5	4	7
5	1	6	3	2	3	5	6	1	6
2	4	3	1	5	4	2	7	3	2

ISBN 9780170413534

a Complete a frequency table for this data.

b Draw a combined frequency histogram and polygon to display this data.

> **Frequency polygon** is a line graph formed by joining the midpoints of the tops of the columns of a **frequency histogram**.

c On how many days did Hadieya make less than three calls?

3 This data shows the number of accidents on the M5 motorway each day for a two week period in September.

1 0 0 2 0 3 5 1 0 1 0 2 3 0

a Summarise this data in a frequency table.

b Draw a dot plot for this data.

c Is it easier to understand this data in the table or in the graph? Give a reason for your answer.

4 The Student Representative Council at Nelson Valley High School wanted to gather data on the number of students using the school canteen. They recorded the number of students served daily at the canteen over a three week period.

105   76   97   88   114   86   114   101

112   98   95   105   117   81   112

a Summarise this data in a grouped frequency table using classes 70–79, 80–89, … 110–19.

b Draw a stem-and-leaf plot for this data.

c Is it easier to understand this data in the table or in the graph? Give a reason for your answer.

13.06 Measures of central tendency

The next step in the statistical investigation process is to analyse the data we have collected and organised.

In Year 11, we calculated three **measures of central tendency** which are 'typical values' that represent the centre or average of the data.

> The **mean** is calculated by adding all the scores and dividing by the number of scores. This is what most people call the 'average'.
>
> The **median** is the middle score when the scores are arranged in order from smallest to largest. If there is an even number of scores, it is the average of the 2 middle scores.
>
> The **mode** is the most common or frequent score(s).

Scoring statistics

Statistical measures

EXAMPLE 3

Gemmy investigated waiting times for patients at a hospital by timing how long it took from the patient's arrival until they were seen by a doctor. This **stem-and-leaf plot** shows the results in minutes.

Stem	Leaf
0	2 9
1	7 8 9
2	0 2 4 7
3	4 4 4 8 9
4	2 3 5 5 7
5	2 4 8 8 9

For the set of waiting times, find the:

a mean   **b** median   **c** mode.

Solution

a To calculate the mean, add all the scores and divide by the number of scores. In this example, there are 24 scores.

Sum of the scores $= 2 + 9 + 17 + 18 + 19 + 20 + 22 + ... + 58 + 58 + 59$
$$= 840$$

$$\text{Mean} = \frac{840}{24}$$
$$= 35$$

Mean waiting time was 35 minutes.

b Because there are 24 scores (an even number), the median will be the average of the 12th and the 13th scores. The 12th score is the last of the 34s and the 13th score is 38 (circled in red).
The median will be halfway between 34 and 38.

Stem	Leaf
0	2 9
1	7 8 9
2	0 2 4 7
3	4 4 ④ ⑧ 9
4	2 3 5 5 7
5	2 4 8 8 9

$$\text{Median} = \frac{34 + 38}{2}$$
$$= 36$$

c The mode is the most common score.   Mode $= 34$

Hint: To remember that the mode is the most common score, note that both 'mode' and 'most' start with 'mo'.

Exercise 13.06 Measures of central tendency

1 Carol's personal trainer recorded her pulse every 2 minutes while she was working out on a cross-country machine at the gym. The stem-and-leaf plot shows the results.

Example
3

Stem	Leaf
12	7 9
13	0 3 5
14	3 3 3 6
15	4 5 7

a How many times did the personal trainer record Carol's pulse?

b How long did Carol work out on the cross-country machine?

c Calculate Carol's mean pulse.

d What was her median pulse?

e Find the mode of the data.

2 Carol noticed that the gym's 20 walking machines are very popular. She recorded the number of walking machines that were vacant each day when she arrived at the gym. She used a dot plot to show the information.

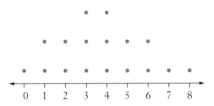

a What is the modal number of vacant walking machines?

b Calculate, correct to one decimal place, the mean number of vacant walking machines.

c Calculate the median number of vacant machines.

d In what percentage of times were fewer than 4 walking machines vacant?

e When Carol arrives at the gym tomorrow, based on her data, what is the probability that there won't be any vacant walking machines for her to use?

3 Carol is trying to improve the tone of her upper arms. Each day, she counted the number of bicep curls she completed in 30 seconds and displayed the data in a dot plot.

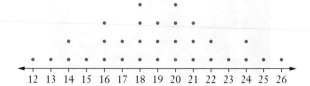

a How many times did Carol count the number of bicep curls she completed in 30 seconds?

b What was the mode?

c Determine the median.

d Calculate the mean.

4 Carol uses the leg extension machine to tone the muscles in her legs. The bar chart shows information about the number of repeats she is able to complete on the leg extension machine.

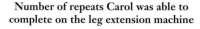

Number of repeats Carol was able to complete on the leg extension machine

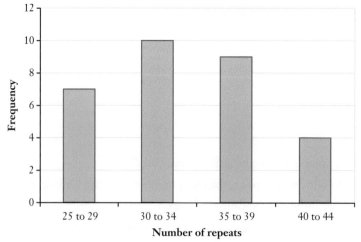

a What is the modal class?

b We can't determine the exact value of the median from the graph. Explain how you know that the median is in the class 30 to 34.

c Copy and complete the frequency table from the graph. The first row has been completed.

Class centre is the centre of a **class interval**.

Number of repeats	Class centre, cc	Frequency, f	$f \times cc$
25 to 29	27	7	189
30 to 34			
35 to 39			
40 to 44			
Totals			

d Use the table to calculate an estimate for the mean, correct to the nearest whole number.

5 At the end of each gym session, Carol recorded the number of seconds it took her pulse to fall to less than 100 beats/minute. She used a bar chart to display the information.

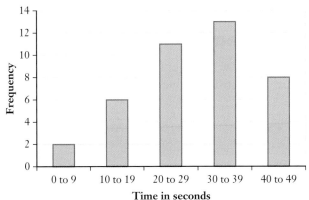

a What is the modal class?

b How many gym sessions are represented on the graph?

c Where do you think the median is?

d Use a calculation to explain how you know that the class centre of the class 0 to 9 is 4.5.

e Use a table with an $f \times cc$ column to estimate the mean number of minutes after exercise before Carol's pulse fell to less than 100 beats/minute.

ARE SCHOOL BACKPACKS GETTING TOO HEAVY?

Parents and students often complain about the weight of school books and backpacks. In this activity, we are going to weigh backpacks that are loaded with school books and equipment.

You will need a set of scales and permission to weigh backpacks that are located outside a classroom.

1 Measure and record the mass of each backpack, correct to the nearest kg.

2 Construct a stem-and-leaf plot or a dot plot to display the data.

3 Calculate the mean, median and mode.

4 The backpacks you weighed came from outside a classroom. Explain why the weights of backpacks when students are carrying them will be more than the weights you have measured.

5 Medical experts believe that a school backpack should weigh less than 20% of the weight of the person carrying it. How could you determine whether the backpacks you weighed are an appropriate weight for the students who carry them?

6 If you discover that the backpacks are too heavy, what could you do about it?

13.07 Measures of spread

Statistical measures

Another way of analysing data is to consider how spread out it is. In Year 11, we learned about three **measures of spread**, which are statistical values that describe how the data is spread out.

Statistical calculations

> The **range** is the difference between the highest score and the lowest score.
>
> The **interquartile range** is the difference between the **upper quartile** Q_3 and the **lower quartile** Q_1.
>
> The **standard deviation** describes how far each data score 'deviates' (is different) from the mean. It has the symbol σ_n or σ, which is the Greek letter 'sigma'.

Statistics review

The more spread out the data, the bigger the range, the interquartile range and the standard deviation.

Interquartile range

Statistical measures puzzle

EXAMPLE 4

A parent group is concerned about the amount of alcohol and smoking shown in films. The parents watched 20 current movies and recorded the number of minutes they saw alcohol and smoking in the film. They presented the data in separate **box plots**.

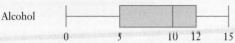

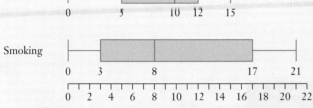

a Which measure of central tendency is shown on a box plot?

b Calculate the range of times for alcohol and smoking in the films.

c Calculate the interquartile ranges for alcohol and smoking.

d Which graph has the greater standard deviation?

Solution

a

The median is shown on a box plot. It is indicated by the line at the centre of the box.

b Range = highest score – lowest score.

Range for alcohol = 15 – 0
$$= 15$$
Range for smoking = 21 – 0
$$= 21$$

c The edges of the box plot mark the upper and lower quartiles, so the length of the box is the interquartile range.

Interquartile range for alcohol = 12 – 5
$$= 7$$
Interquartile range for smoking = 17 – 3
$$= 14$$

d The data for smoking is more spread out than the data for alcohol.

The graph for smoking has the larger spread and hence the larger standard deviation.

EXAMPLE 5

This **dot plot** shows the number of goals scored by a soccer team per match over 11 matches.

a What is the range of goals scored per match?

b Calculate the interquartile range.

c Calculate the standard deviation, correct to one decimal place.

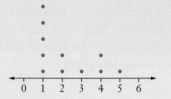

Solution

a

> Be careful! A common error is using the 6 and the 0 at the end of the dot plot to calculate the range. We don't use these numbers because there aren't any scores of 6 or 0.

Range $= 5 - 1$
$= 4$

b Place the data in order. Find the middle score (median) and then the middle of each half of the scores.

$1 \quad 1 \quad ①\quad 1 \quad 1 \quad ②\quad 2 \quad 3 \quad ④\quad 4 \quad 5$

$\quad\quad\quad Q_1 = 1 \quad\quad\quad Q_2 = 2 \quad\quad\quad Q_3 = 4$

Upper quartile   Median   Lower quartile

Interquartile range $= Q_3 - Q_1$
$= 4 - 1$
$= 3$

c To calculate standard deviation, follow the instructions for the statistics mode of your calculator as shown in the table below.

Standard deviation $\sigma_n = 1.4200\ldots$
≈ 1.4

Operation	Casio scientific	Sharp scientific
Start statistics mode.	MODE STAT 1-VAR	MODE STAT =
Clear the statistical memory.	SHIFT 1 Edit, Del-A	2ndF DEL
Enter data.	SHIFT 1 Data to get table 1 = 1 = etc. to enter in column AC to leave table	1 M+ 1 M+ etc.
Calculate the population standard deviation. ($\sigma_n \approx 1.42$)	SHIFT 1 Var $x\sigma_n$ =	RCL σx
Return to normal (COMP) mode.	MODE COMP	MODE 0

Exercise 13.07 Measures of spread

1 Each day, for one month, Darren recorded the price of E10 petrol (in cents/litre) at his local garage and at a freeway service station. He displayed the data in a pair of box plots.

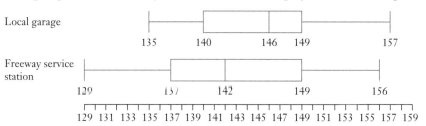

a What was the median price at each location?

b Calculate the price range at each location.

c Find the interquartile range at each location.

d Which box plot shows a smaller standard deviation? Justify your answer.

2 Paul's golf scores for 15 rounds of golf are shown on this dot plot.

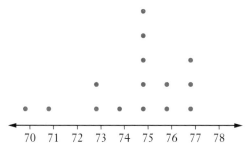

a What was Paul's modal score?

b What is the range of Paul's scores?

c When there are 15 scores arranged in order, the 4th score is the lower quartile, the 8th score is the median and the 12th score is the upper quartile. What is the median?

d Find the lower and upper quartiles.

e Calculate the interquartile range.

f In golf, low scores are better than high scores and 'par' (the target score) at Paul's golf course is 72. Based on the data in this question, what is the probability that in his next game of golf Paul's score will be below par?

3 This stem-and-leaf plot from p. 401 shows Carol's pulse rate every 2 minutes while she was exercising on a cross-country machine at the gym.

Stem	Leaf
12	7 9
13	0 3 5
14	3 3 3 6
15	4 5 7

  a Calculate the range of Carol's pulse rate.

  b Calculate the standard deviation, correct to one decimal place.

  c Find the lower and upper quartiles.

  d Calculate the interquartile range.

  e How many of Carol's pulse rates were higher than the lower quartile but lower than the upper quartile?

  f Calculate the percentage of Carol's pulse rates that were higher than the lower quartile but lower than the upper quartile.

  g Is your answer to part **f** what you expected? Why or why not?

4 Driving at unsafe speeds is a community problem. These histograms and box plots show the ages of male and female drivers caught speeding on a freeway during a holiday weekend.

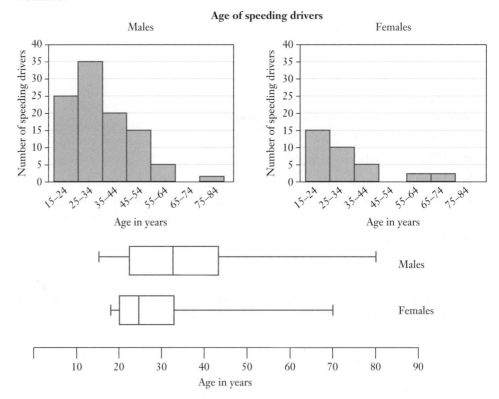

Age of speeding drivers

  a How old was the youngest person caught speeding?

  b How old was the oldest female caught speeding?

ISBN 9780170413534

c What is the modal class for:

    i males?   **ii** females?

d Find the median female age for speeding.

e Calculate an estimate for the mean male age for speeding, correct to the nearest year.

f Find the range in the ages of the males caught speeding.

g What is the interquartile range of the female drivers caught speeding?

h Which group, males or females, has the larger standard deviation? Justify your answer.

INVESTIGATION

COMPARING THE SIZES OF MATHS BOOKS WITH ENGLISH NOVELS

1 Ask your maths teacher to bring a copy of every maths textbook on their desk in the staffroom and your English teacher to lend you a copy of every novel used in senior English classes.

2 Record the number of pages in each book, then construct back-to-back stem-and-leaf plots or side-by-side dot plots to compare the number of pages in each maths text and in each English novel.

3 Calculate the mean number of pages for each type of book.

4 Calculate the standard deviation for the number of pages for each type of book.

5 'Maths textbooks have more pages than English novels, but there is a greater variation in the number of pages in English novels.' Do you agree with this statement? Does your data support this? Justify your answer.

13.08 Making conclusions

When you have completed all the stages of a statistical investigation, it is time to interpret your findings and present your conclusions in a format appropriate to the situation – written report, PowerPoint presentation or a poster/wall chart. It is important to make your presentation easy to understand.

Exercise 13.08 Making conclusions

In this exercise, you will be given the findings of different investigations and asked to interpret them. When you have completed the exercise, in a small group, share your answers with other students.

---

1 Andrew, the market researcher from Exercise 13.02 on p. 394, summarised the findings for attendance at different types of entertainment provided by the United Club in the table below. It shows the percentage of members surveyed who answered in each category.

	Never	Once a year	2–3 times per year	Once a month	Weekly
Concerts	3%	22%	57%	11%	7%
Films	2%	5%	32%	44%	17%
Dances	39%	37%	18%	6%	0%
Trips away	18%	28%	19%	20%	15%

Write a short paragraph explaining what recommendations Andrew should make about:

a the types of entertainment offered by the club

b the frequency of entertainment offered by the club.

2 Andrew also asked a question about how often people ate at the bistro. These are his results.

Never	Once a year	2–3 times per year	Once a month	Weekly	2–3 times per week
3%	8%	24%	31%	26%	8%

Write a short paragraph explaining what recommendations Andrew should make based on this data. Include suggestions about what further information would help him to decide what changes are needed at the bistro.

3 Janelle surveyed Year 12 students about the number of hours of paid work they completed in one week. When she analysed her data, she found the following.

Mean number of hours = 6   Standard deviation = 3.54

Modal number of hours = 3   Range = 12

Median number of hours = 4   Lower quartile = 2

Upper quartile = 6   Interquartile range = 4

Least hours worked = 0   Highest hours worked = 12

The principal of Janelle's school is concerned that Year 12 students are doing too much paid work and that this is having a negative effect on their studies. Do these results support that view? Write a short paragraph about these results to support your answer and draw a box plot for the data.

4 In Exercise 13.05, Question **2** on p. 398, Hadieya recorded the number of phone calls she made each day for the month of June. For this data:

Mean = 3.8   Median = 3.5   Mode = 3   Range = 6

Optel offers Hadieya a plan where the first 10 phone calls each day are free. After that, calls are charged at $2 per minute. Should she take the plan or not? Justify your answer using this data.

5 In Exercise 13.07, Question **1** on p. 407, Darren recorded the price of petrol at his local garage and at a freeway service station over one month. He found the following:

Local garage: Median = 146c/L   Range = 22c   Interquartile range = 9c

Freeway: Median = 142c/L   Range = 27c   Interquartile range = 12c

Darren's friends have asked him to recommend a cheap location for fuel. Write a short paragraph explaining which service station Darren should recommend to his friends. Justify your answer. You may wish to look back to the box plots shown in Exercise 13.07, Question **1**

6 In Exercise 13.07, Question **4** on p. 408, you were given information about the ages of male and female drivers caught speeding on a freeway during a holiday weekend. This data showed the following:

Number of males = 100	Number of females = 34
Median age for males = 32	Median age for females = 24
Range for males = 45	Range for females = 52
Interquartile range for males = 20	Interquartile range for females = 12

You are asked to make recommendations for an advertising campaign to reduce the number of people speeding on holiday weekends. Based on this data, write a short paragraph with your recommendations. You may like to refer to the data provided in Exercise 13.07, Question **4**.

INVESTIGATION

TRAVELLING TO SCHOOL

Part A

In this investigation, you are going to use what you have learned in this chapter.

1 Posing the question

You are going to conduct a survey of how students in your school travel to school and how long it takes them.

2 Collecting data

Before you start, you will need to make some decisions.

Who will you ask?

- How large a sample will you use?
- How will you select your sample?

The questionnaire

- What questions will you ask?
- What are the possible categories or types of travel that students at your school might use? List them.

- What will you do if there is a category you haven't thought of?
- What will you do if someone uses more than one method of transport?
- How will you record the time taken to travel to school?
- Check for any bias.

Collect the information

3 Organising and representing the data

- Organise your data into frequency tables OR use a spreadsheet to organise your data.
- Display your data using appropriate graphs.

4 Analyse the data

- Where appropriate, find the mean, median and/or mode.
- Find the range, interquartile range or standard deviation as required.

Answer the following questions.

- What is the most common method of travel used?
- What is the least common method used?
- What is the longest time taken to travel to school?
- Which measure of central tendency (mean, median or mode) is most appropriate for your time data? Justify your choice.
- What effect does the location of your school have on the method of travel used by students and the time taken to travel to school?

5 Communicating your results

Present your results in a report – this may be written or on PowerPoint.

- Suggest reasons for the collection of this data.
- Summarise the information you have found.
- Consider the impact of changing the start time of your school to 15 minutes earlier. Make recommendations about this in light of the data you have collected.

Part B

Choose your own topic/question and complete a statistical investigation. When writing the report, show how you have completed each of the steps.

ISBN 9780170413534

WORD MATCH

Match each term to its definition.

Revision
crossword

1	Bias	**A**	A common way to collect information
2	Census	**B**	An unwanted influence that makes results unreliable
3	Frequency tables	**C**	The group of people to ask for the information we need
4	Graphs	**D**	When we survey every person in the population
5	Measures of central tendency	**E**	When we survey a small group out of the whole population
6	Measures of spread	**F**	A way of organising data
7	Questionnaire	**G**	Visual representations of data
8	Recommendations	**H**	Typical values that represent the average of the data – mean, median, mode
9	Sample	**I**	Values that describe how the data is spread out – range, interquartile range, standard deviation
10	Target population	**J**	Suggestions for action as a result of the findings of a statistical investigation

SOLUTION ^{TO}_{THE} CHAPTER PROBLEM

Problem

Avril is employed by a large car manufacturer to research the available optional extras that customers want in a new car. How is she going to do this?

Solution

Avril needs to complete a statistical investigation.

1 She needs to decide what optional extras could be made available, such as wheel rims, different colours and so on.

2 She needs to decide on the target population – who she will ask. This could be previous customers or adults aged over 17 or adults above a certain income level.

3 She should then create a questionnaire, choose her sample and collect her data.

4 Once she has her data, she needs to organise it and create graphs to illustrate the important points.

5 She needs to analyse her data using appropriate measures. This would depend on what questions she asked.

6 She would then need to write a report presenting her conclusions in a way that can be easily understood. She may also be required to make a presentation to the executives of the car manufacturer.

- What have you learned that's new in this topic?
- Give examples of organisations or groups that would carry out statistical investigations.
- Is there anything you didn't understand? If so, ask your teacher for help.

Copy and complete this mind map of the topic, adding detail to its branches and using pictures, symbols and colour where needed. Ask your teacher to check your work.

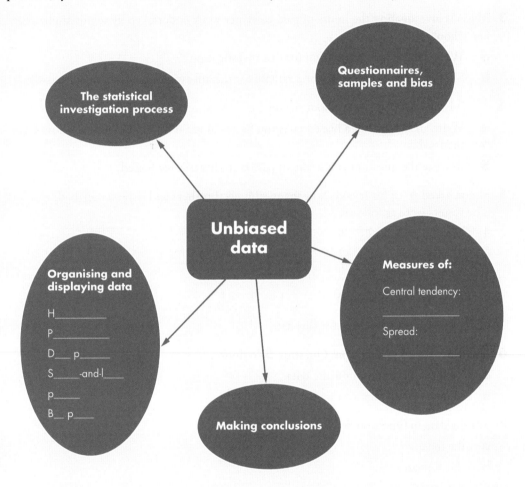

The statistical investigation process

Questionnaires, samples and bias

Unbiased data

Organising and displaying data

H_____
P_____
D__ p_____
S_____-and-l____
p_____
B__ p____

Making conclusions

Measures of:

Central tendency:

_____

Spread:

_____

13. TEST YOURSELF

 Example 13.01

1 In your own words, outline the steps involved in a statistical investigation.

 Example 13.02

2 Anthony wrote the following question as part of a questionnaire he was giving to Year 12 students about their consumption of alcohol. 'How often do you drink?'

   a What problems might people have in answering this question?

   b Rewrite the question, providing some tick box options.

 Example 13.03

3 Maria is investigating the hours of paid work per week undertaken by senior students at her school.

   a Who is the target population for this investigation?

   b Should Maria use a census or a sample for this investigation? Justify your answer.

 Example 13.04

4 Tranh is surveying students about the quality of food sold at the school canteen.

   a Write an example of a biased question he could use if he wanted to show how great the food is.

   b Rewrite the question you wrote in part **a** so that it is *not* biased.

 Example 13.05

5 Simon asked 40 of his friends how many siblings (brothers and sisters) they had. The results were as follows.

4	0	2	3	2	4	3	5	4	3
0	0	1	2	3	2	2	1	4	3
1	1	1	5	2	5	0	1	2	1
0	1	3	2	3	4	1	1	3	3

Complete a frequency table for this data.

 Example 13.05

6 Use the frequency table from Question **5** to draw a:

   a frequency histogram for the data

   b frequency polygon for the data.

 Example 13.06

7 For the data in Question **5** find:

   a the mean

   b the median

   c the mode.

NCM 12. Mathematics Standard 1   ISBN 9780170413534

8 For the data in Question **5** find:

   a   the range

   b   the interquartile range

   c   the standard deviation (correct to 2 decimal places).

Example
13.07

9 Andrew, our market researcher, asked a question about the cost of entertainment events at the United Club. These are his results:

Example
13.08

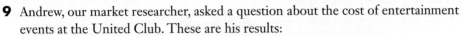

Cheap	Reasonable	Expensive	Don't attend events as they are too expensive
10%	45%	22%	23%

Write a short paragraph explaining what recommendations Andrew should make based on this data.

Practice set 3

Section A: Multiple-choice questions

For each question, select the correct answer **A**, **B**, **C** or **D**.

Exercise
10.01

1 During exercise, Robbie had an average pulse rate of 140 beats/minute.
How many times did her heart beat during her 40-minute exercise session?

   A   3.5   **B**   210   **C**   5600   **D**   8400

Exercise
11.01

2 Find the length of side h in this diagram.

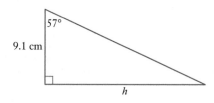

   A   5.0 cm   **B**   5.9 cm   **C**   7.6 cm   **D**   14.0 cm

Exercise
13.01

3 Joshua creates 3 graphs showing the information he found in his survey.
What step is this in the statistical investigation process?

   A   collecting data   **B**   organising data

   C   representing data   **D**   analysing data

Exercise
12.01

4 These 2 shapes are similar. Which statement is true?

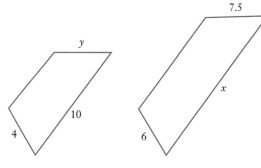

   A $\dfrac{x}{10} = \dfrac{y}{7.5}$ **B** $\dfrac{x}{10} = \dfrac{6}{4}$ **C** $\dfrac{y}{7.5} = \dfrac{6}{4}$ **D** $\dfrac{x}{6} = \dfrac{y}{4}$

NCM 12. Mathematics Standard 1   ISBN 9780170413534

5 James is 25 years old and he exercises at an intensity such that his pulse rate is 80% of his maximum heart rate. Calculate James' exercise pulse rate given the formulas for MHR on page 422.

Exercise 10.02

   A   156 beats per minute   **B**   161 beats per minute

   C   195 beats per minute   **D**   201 beats per minute

6 This diagram shows a stretch of road. The road rises 92 m for every 500 m travelled along the road. We want to find the angle at which the road is inclined to the horizontal. Which trigonometric ratio will we use?

Exercise 11.02

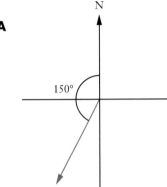

   A   sine   **B**   cosine   **C**   tangent

7 Which diagram shows a bearing of 150°?

Exercise 11.05

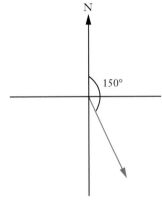

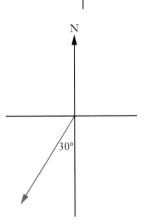

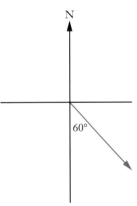

8 Evan is using a map with a scale of 1: 20 000. The distance to the shops on the map is 3.4 cm. How far will Evan have to walk to the shops?

Exercise 12.02

   A   57 m   **B**   680 m   **C**   571 m   **D**   6800 m

Exercise 13.06

9 Competitors' scores in a diving competition are shown below.

6.2 7.0 8.2 8.8 7.0 9.1 7.0 8.1 8.0

What is the median score?

A 2.9 **B** 7.0 **C** 7.7 **D** 8.0

Exercise 13.07

10 Calculate the interquartile range of the data in Question **9**.

A 1.6 **B** 1.2 **C** 1.9 **D** 2.9

Exercise 11.03

11 Meredith is flying a glider at an altitude of 2 km. To land, she descends at an angle of depression of 20°. Which diagram illustrates this correctly?

A

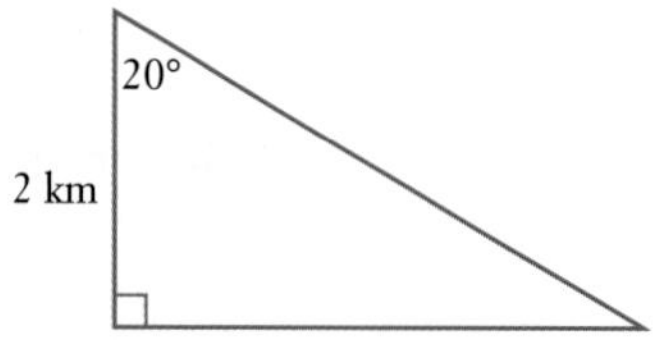

B

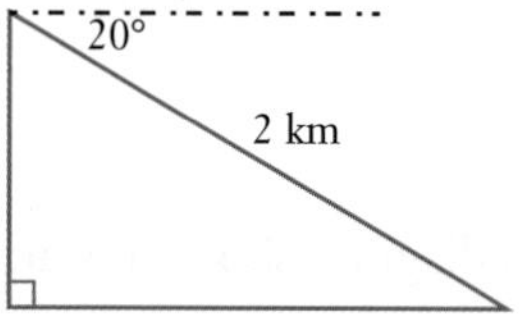

C

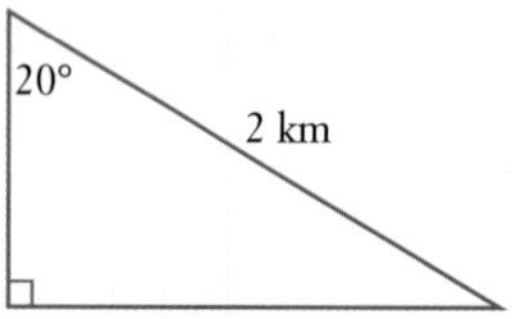

D

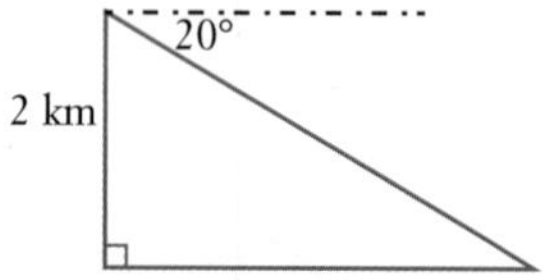

Exercise 12.03

12 Which diagram shows the northern elevation for the garage shown on the plan?

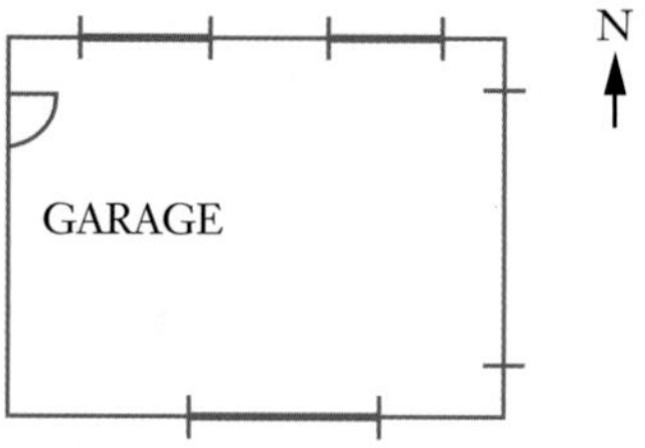

A

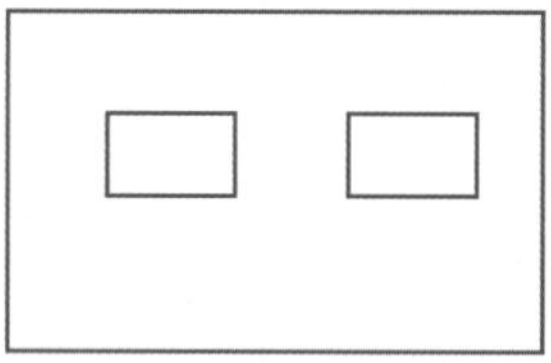

B

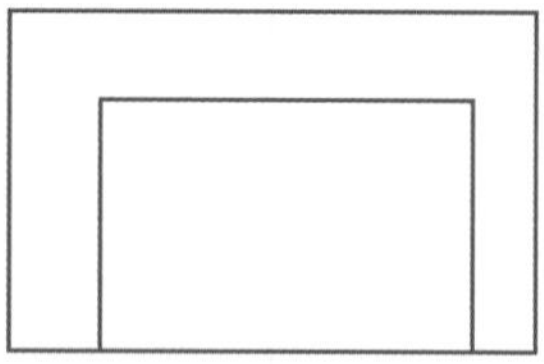

C

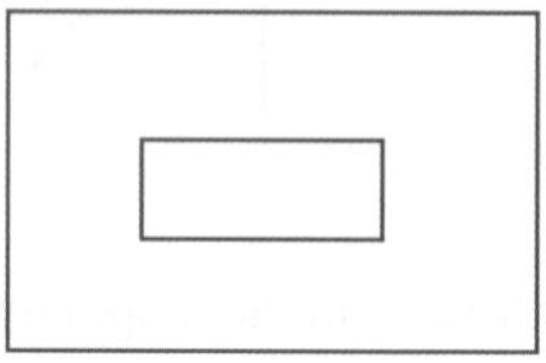

D

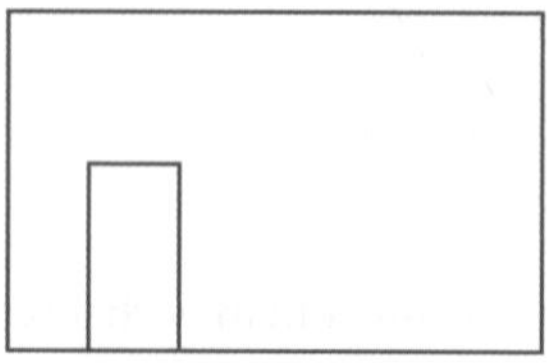

ISBN 9780170413534

Section B: Short-answer questions

1 Malek is exercising and he takes his pulse. He counts 39 beats in 15 seconds. Calculate his pulse in beats per minute.

Exercise
10.01

2 Calculate the value of the pronumeral in each diagram.

Exercise
11.01

a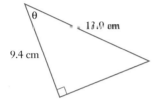

b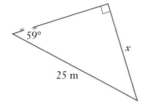

Answer to the nearest minute.   Answer correct to two decimal places.

3 a Jim's survey asks 'How often do you go out to dinner?' Suggest a series of tick boxes that Jim could provide for people to record their answers.

Exercise
13.02

b Jim's survey also asked the question: 'Rate your meal at the fabulous Diggies' Steak House':

Exercise
13.04

Great   Yummy   Quite nice

i In what way is this question biased?

ii Rewrite it so it is not biased.

4 In each pair of similar figures, find the scale factor from the left shape to the right shape and find the value of each pronumeral.

Exercise
12.01

a

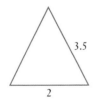   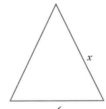

b

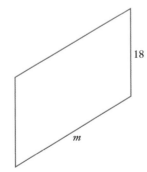   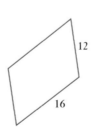

ISBN 9780170413534

5 Lily is 24 years old and goes to the gym 3 times a week for aerobic exercise. She has a resting pulse of 64 beats/minute.

Maximum heart rates (MHR)	Target heart rate
MHR for males = 220 − age in years	THR = $I \times$ (MHR − RHR) + RHR where:
MHR for females = 226 − age in years	RHR = resting heart rate
	MHR = maximum heart rate
	I = exercise intensity as a decimal

Recovery or weight loss	Aerobic	Anaerobic
60% to 70%	70% to 80%	80% to 90%

Use the formulas and information given to calculate:

a Lily's maximum heart rate

b Lily's minimum target heart rate for exercise

c Lily's maximum target heart rate for exercise

6 Measure the length of each scale drawing, then use the scale ratio to calculate its actual length.

a

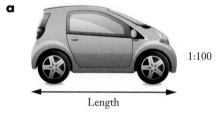

1:100

Length

b

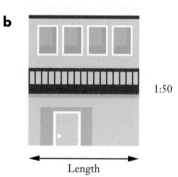

1:50

Length

7 a Write the compass bearing from O to W.

b Write the three-figure (true) bearing from O to W.

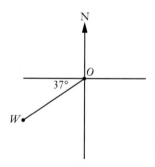

8 These marks out of 50 are for an assessment task of 2 Year 11 classes. Exercise 13.05

48	19	17	45	39	27	40	41	30	23	38
32	30	27	31	34	36	20	25	22	40	41
30	46	27	34	31	23	8	38			

  a Summarise this data in a grouped frequency table using classes 5–9, 10–14, 15–19, ..., 45–49.

  b Draw a combined frequency histogram and polygon to display this data.

  c How many students scored less than 30 out of 50?

9 Use the data and your answers from Question **8** for these questions. Exercise 13.06

  a What is the modal class for the grouped data?

  b Calculate the mean test score. Answer correct to one decimal place. Exercise 13.07

  c What is the range of test scores?

  d Find the median of the test scores.

  e Find the interquartile range for these test scores.

10 Linda is standing 750 m from the base of a tree. The angle of elevation to the top of the tree is 7°. Exercise 11.03

  a Draw a diagram showing this information.

  b Find the height of the tree, correct to one decimal place.

11 Brayden's blood pressure is 112/77. Exercise 10.03

  a What is his systolic pressure?

  b What is the pressure in Brayden's arteries when his heart isn't pumping?

  c Does he have normal blood pressure?

12 From the top of a cliff, the angle of depression to a weather buoy is 15°. Exercise 11.03

The buoy is moored 420 m from the cliff.

Find the height of the cliff, correct to one decimal place.

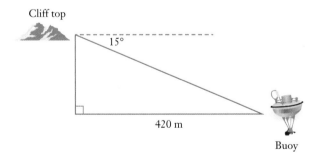

13 Tim and Kim buy a house with the floor plan given in Exercise 12.04, Question **2** on page 365.

a What are the dimensions of the garage?

b Tim and Kim decide to concrete the garage floor.

   i How many square metres of concrete will they need?

   ii Concrete including delivery costs $75/square metre. How much will the garage floor cost?

c How many windows are in this house?

d The carpet in bedroom 1 needs to be replaced.

   i How many square metres of carpet will be needed?

   ii Carpet costs approximately $85/m^2. How much will it cost to replace the carpet?

e The ceilings in all 3 bedrooms need to be repainted.

   i How many square metres of ceiling need to be repainted?

   ii One litre of ceiling paint covers 12 m^2. White ceiling paint costs $49.90 for a 4 L tin. Calculate the cost of repainting the ceilings if 2 coats of paint are needed.

14 For the frequency table below, find:

a the mean of the data

b the standard deviation of the data, correct to 2 decimal places.

Average hours of sleep	Frequency
5	6
6	4
7	9
8	13
9	4
10	4

15 Coffs Harbour is on a bearing of 024° from Sydney and is 435 km from Sydney. Calculate, correct to the nearest kilometre, how far:

a east of Sydney is Coffs Harbour

b south of Coffs Harbour is Sydney

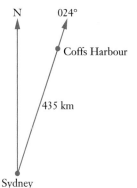

Exercise
11.06

16 This is the design for a built-in wardrobe. All measurements are in millimetres. Draw a scale drawing of this design using a scale of 1 : 20.

Exercise
12.06

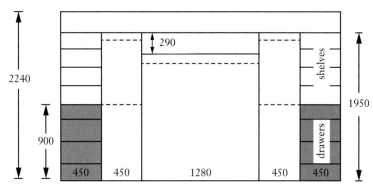

Rails for hanging clothes (- - -). Top rails are 150 mm below the shelves above. Design is symmetrical.

ANSWERS

Chapter 1

Exercise 1.01

1 a $6.00   b $12.15
  c $9.00   d $38.25
  e $69.50   f $4.38
  g $4.19   h $14.84
2 a $756   b $7826
  c $27.60   d $697.51
3 $81.40   4 13.6%   5 10.0%
6 $33.25/h   7 12.5%   8 $8.91
9 35%   10 12.8%   11 $2081
12 a $22.13   b 53%

Exercise 1.02

1 a $115.20   b $243.36
  c $866.25   d $350.63
2 a $44.10   b $193.60
  c $54.27   d $4.58
3 0.37%
4 a 0.6%   b 0.138%
  c 3.6%   d 0.0197%
  e 0.277%   f 1.8%
5 a $2484   b $7084   c 35%
6 a $15.13   b $840.13
7 $141 875
8 a 4   b 8%   c $3000
9 a $1400   b 10.1%
10 10.0%

Exercise 1.03

1

	Interest	Balance
End of the 1st year	$I = Prn$ $= \$10\,000 \times 0.05 \times 1$ $= \$500$	$\$10\,000 + \$500$ $= \$10\,500$
End of the 2nd year	$I = Prn$ $= \$10\,500 \times 0.05 \times 1$ $= \$525$	$\$10\,500 + \$525$ $= \$11\,025$
End of the 3rd year	$I = Prn$ $= \$11\,025 \times 0.05 \times 1$ $= \$551.25$	$\$11\,025 + \$551.25$ $= \$11\,576.25$

2 a $450   b $9450
  c $9922.50   d $10 418.63
  e $1418.63   f 14.4 years
3 a $7600.35   b $600.35
  c Simple interest = $294 per year.
    $294 × 2 = $588 simple interest for 2 years.
    $600.35 − $588 = $12.35 less interest over
    the 2 years.
4 a

	Interest	Balance
End of the 1st month	$I = Prn$ $= \$12\,000 \times 0.006 \times 1$ $= \$72$	$\$12\,000 + \$72$ $= \$12\,072$
End of the 2nd month	$I = Prn$ $= \$12\,072 \times 0.006 \times 1$ $= \$72.43$	$\$12\,072 + \$72.43$ $= \$12\,144.43$
End of the 3rd month	$I = Prn$ $= \$12\,144.43 \times 0.006 \times 1$ $= \$72.87$	$\$12\,144.43 + \$72.87$ $= \$12\,217.30$

  b $217.30
5 a $9600   b 0.8%
  c 9.6%   d $154.21

6 a $5208.75

b Scam indicators include:

- cold phone call from someone Zac doesn't know
- push from the 'advisor' to act quickly
- very high interest rate
- claim that the opportunity is 'risk-free'
- minimum investment of $10 000.

7 a Cell B8 =B4, Cell D8 =B8+C8, Cell B9 =D8

b We use $ signs for an absolute cell reference. This means that the cell reference stays the same when we use the formula in, for example, a 'fill down'.

c

	A	B	C	D
1	Compound interest spreadsheet			
2	Only enter values in the blue shaded cells			
3				
4	Principal	$1,000.00	Annual rate of interest as a percentage	12%
5				
6				
7		Account balance at the beginning of the year	Interest earned during the year	Account balance at the end of the year
8	Year 1	$1,000.00	$120.00	$1,120.00
9	Year 2	$1,120.00	$134.40	$1,254.40
10	Year 3	$1,254.40	$150.53	$1,404.93
11	Year 4	$1,404.93	$168.59	$1,573.52
12	Year 5	$1,573.52	$188.82	$1,762.34
13	Year 6	$1,762.34	$211.48	$1,973.82
14	Year 7	$1,973.82	$236.86	$2,210.68
15	Year 8	$2,210.68	$265.28	$2,475.96

d $2047.91

Exercise 1.04

1 10 910.78

2 a $PV = 1870, r = 0.068, n = 4$

b $2432.91 **c** $562.91

3 $6461.81

4 a $6.36 \div 12 = 0.53$ **b** $2144.65

c $194.65

5 a 0.49% **b** 60

c $49 623.52

6 a $17 786.85 **b** $1463.63

c $35 607.18 **d** $98 875.58

e $16 897.39

7 $19 726.26 **8** $3200 **9** $8358.32

10 a 7.72% **b** 10.09% **c** 6.05% **d** 7.57%

11 7.72%

12 a 22.5% p.a.

b The rate is very high, more than 20% p.a. Most likely it is a scam and not a safe investment.

13 a $6550

b Months in 3 years = 36, Initial investment = PV = $5600 and the future value = FV = $6550

c 0.44% per month

14 A $11 800 **B** $11 876.48

C $11 888.45; option C is the best

15 Simple interest = $64 000, compound interest = $64 032.10. Compound interest is better.

16 a interest

b Needs brackets = G4*(1+G6/G5)^(G7*G5)

c Teacher to check

Exercise 1.05

1 a $1966.81 **b** $517.13

c $359.36 **d** $366.16

2 a $7193.75 **b** $1393.75

3 $1744.93 **4** $4.19 **5** $464

6 6.4% p.a. monthly compounding gives the better return with $61 274.92 interest. The annually compounding investment gives only $60 880.61 in interest.

7 The Managed funds investment gives the better return; interest = $12 366.26 with $803.81 in fees, giving a return of $11 562.45. The Finance company returns $10 232.38 less $120 in fees, giving $10 112.38.

Exercise 1.06

1 a $20 655 **b** 20 to 21 years

2 $840

3 a $18.13 **b** $24.58

4 a The formula is $FV = PV(1 + r)^n$, where $FV = 38, r = 0.034, n = 10$ years (i.e. 2016 − 2006 = 10), so $38 = PV(1 + 0.034)^{10}$, which can be rearranged to $PV \times (1.034)^{10} = 38$.

b $27.20 **c** $12.27

5 $25.21

6 a 14.4 years

b 14.4, 28.8, 43.2, 57.6, 72

c Almost 50 years ago, approximately 1970

7 24.56, 75.40, 231.48, 710.63, 2181.63

8 a 1410 bolivar

b The actual value is approximately 29 510 bolivars.

9 a $173 644   **b** $23 153
    c $20 837   **d** $775 609
    e $2894   **f** $57 881

Test yourself 1

1 a $35   **b** 58%
2 $125   **3** 4%
4

	Interest	Balance
End of the 1st year	$I = Prn$ $= \$4000 \times 0.05 \times 1$ $= \$200$	$4000 + $200 = $4200
End of the 2nd year	$I = Prn$ $= \$4200 \times 0.05 \times 1$ $= \$210$	$4200 + $210 = $4410
End of the 3rd year	$I = Prn$ $= \$4410 \times 0.05 \times 1$ $= \$220.50$	$4410 + $220.50 = $4630.50

5 a $7779.24   **b** $1379.24
6 a 1.6%   **b** $2312.45
7 • 4% p.a. monthly compounding
     → $12 707.42
   • 4.1% p.a. annually compounding
     → $12 726.37
   • 4.4% p.a. simple→ $12 640
   4.1% annually compounding is the best investment.
8 $394.15
9 $20.69

Chapter 2

Exercise 2.01

1 a 13 : 25   **b** 17 : 100   **c** 5 : 12
    d 150 : 1   **e** 31 : 51 : 101
2 a 1 : 4   **b** 12 : 1   **c** 2 : 3
    d 3 : 1   **e** 8 : 12 : 1   **f** 1 : 2 : 3
3 a $1
    b   **i** 3 : 2   **ii** 3 : 1
       iii 2 : 1   **iv** 1 : 3
4 a 27 : 29   **b** 19 : 27   **c** 29 : 75
    d 29 : 19   **e** 27 : 29 : 19
5 a 3 : 250   **b** 23 : 30   **c** 37 : 120
    d 10 : 3   **e** 2 : 5   **f** 300 : 173

6 a black to red   **b** purple to yellow
    c green to grey   **d** green to all
    e green to red to black
    f yellow to purple to grey
7 a 4 : 3   **b** 3 : 4   **c** 3 : 2   **d** 2 : 1
    e 3 : 2   **f** 3 : 1   **g** 3 : 2   **h** 6 : 1
    i 5 : 1   **j** 8 : 7   **k** 7 : 13   **l** 4 : 1
    m 2 : 3 : 1   **n** 2 : 1 : 6   **o** 3 : 5 : 10   **p** 3 : 2 : 4
8 a 1 : 2   **b** 2 : 3   **c** 1 : 3   **d** 4 : 1
    e 2 : 3   **f** 1 : 2   **g** 5 : 3   **h** 3 : 1
9 a 3 : 20   **b** 1 : 20   **c** 3 : 1
    d 1 : 8   **e** 3 : 1 : 20
10 a 8 : 3   **b** 7 : 3   **c** 1 : 4
     d 6 : 1   **e** 16 : 7 : 1
11 a 5 : 8   **b** 1 : 13   **c** 6 : 13
     d 9 : 52   **e** 8 : 5 : 3
12 heavy-duty cleaning

Exercise 2.02

1 a 10   **b** 9
2 a zinc 25 g, nickel 5 g
    b copper 112 g, zinc 40 g
    c 160 g
3 a 5 cups   **b** 18 cups
4 a 2000 mL   **b** 2 L   **c** 150 mL
    d 25 mL   **e** general perennial weeds
5 a 60 mL   **b** 100 mL   **c** 75 mL
6 a 120 L   **b** 8 L   **c** 340 mL
7

Plant	N : P : K	Nitrogen (kg)	Phosphorus (kg)	Potassium (kg)	Total mixture (kg)
Vegetables	6 : 4 : 5	54	36	45	135
		42	28	35	105
Fruit trees	7 : 1 : 2	105	15	30	150
		140	20	40	200

8 a water 250 g, yeast 25 g
    b flour 4 kg, yeast 0.25 kg
    c flour 48 kg, water 30 kg
9 a new brick construction
    b 2 buckets
    c lime 2 buckets, cement 2 buckets
    d sand 20 buckets, lime 6 buckets

Exercise 2.03

1 a $840, $1260   **b** $1750, $350
   c $1200, $600, $300   **d** $450, $600, $1050

2 a 171 L, 684 L   **b** 20, 15
   c 10 800 m^2, 13 500 m^2
   d 637 kg, 182 kg
   e 12 min, 16 min, 24 min
   f $1600, $200, $1000

3 1.5 t

4 399 boys

5 silver beads 20, porcelain balls 15, crystal eyedrops 10

6 160 mL

7 a 2 g   **b** gold 315 g, copper 84 g

8 a 150 mL   **b** 400 mL   **c** 200 mL

9 a 3000 mL
   b

Type of crop	Ratio of concentrate to water	mL of concentrate required	mL of water required
Broccoli	1 : 14	200	2800
Cauliflower	1 : 19	150	2850
Leafy vegetables	1 : 9	300	2700

10 dye A 75 mL, dye B 45 mL

11 a 14.4 L   **b** 0.6 L
   c 600 mL   **d** 0.5 L = 500 mL

12 nitrogen 75 kg, phosphorus 15 kg, potassium 30 kg

Exercise 2.04

1 a 30 m   **b** 116 m   **c** 190 m

2 a i 750 m   **ii** 1375 m   **iii** 2375 m
   b i 40 cm   **ii** 2 cm   **iii** 16 cm

3 33 cm   **4** 5.6 cm

5 a 1 : 10 000   **b** 1 km
   c 680 m   **d** 1.4 km   **e** 11:28 a.m.

6 992 m

7 a 200 km   **b** 36 km/h
   c The average speed is low. The road is probably not sealed and is rough in places.

8 a 20 m   **b** 60 m and 115 m
   c $4900

9 a Even numbers from 156 to 190 and 175, 177, 181
   b around 21 m   **c** 95 m   **d** 230 m
   e It will flood during 1-in-50 year floods.

10 a 9.2 km   **b** 2 h 18 min
   c south west of North Era

Test yourself 2

1 a 9 : 5   **b** 6 : 2 = 3 : 1

2 a 5 : 2000   **b** 1 : 400

3 a 2 : 5   **b** 2 : 3   **c** 2 : 1
   d 3 : 2   **e** 3 : 2   **f** 2 : 1
   g 5 : 1   **h** 1 : 4

4 a 1 : 25   **b** 150 : 1   **c** 4 : 1
   d 1 : 2   **e** 5 : 1   **f** 10 : 1
   g 3 : 1   **h** 3 : 2

5 5 mL

6 1.52 m

7 a 18 cm, 24 cm   **b** 72 cm

8 a 14 and 6   **b** 36 and 24
   c 96 and 24   **d** 50 and 25
   e 20 and 30   **f** 48 and 36

9 a red 18 mL, yellow 6 mL
   b red 8 mL, yellow 2 mL, white 2 mL

10 pearls 24, seed beads 12 and crystals 18

11 a i 920 mm   **ii** 640 m
     iii 480 m   **iv** 1000 m
   b i 880 m   **ii** 440 m
   c 360 m   **d** 120 m by 240 m
   e 400 m   **f** 20 cm, teacher to check

Exercise 3.01

1 a i

x	−2	−1	0	1	2
y	−5	−3	−1	1	3

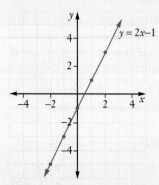

$y = 2x - 1$

ii gradient = 2, y-intercept = −1

b i

x	−2	−1	0	1	2
y	8	6	4	2	0

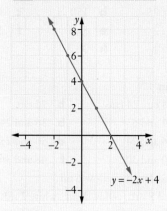

$y = -2x + 4$

ii gradient = −2, y-intercept = 4

c i

x	−2	−1	0	1	2
y	0.5	1	1.5	2	2.5

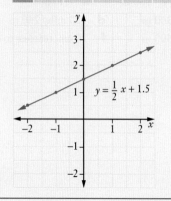

$y = \frac{1}{2}x + 1.5$

ii gradient = 0.5, y-intercept = 1.5

d i

x	−2	−1	0	1	2
y	10	8	6	4	2

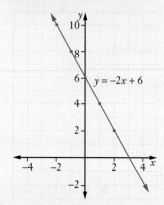

$y = -2x + 6$

ii gradient = −2, y-intercept = 6

2 a i 2   **ii** 1   **iii** $y = 2x + 1$
  b i −2   **ii** 5   **iii** $y = -2x + 5$
  c i 1   **ii** 2   **iii** $y = x + 2$

3 a i gradient = 3   **ii** y-intercept = −5
  b i gradient = −1   **ii** y-intercept = 6
  c i gradient = 2   **ii** y-intercept = 11
  d i gradient = $\frac{1}{4}$ **ii** y-intercept = −5

4 a D   **b** B   **c** C   **d** A

Exercise 3.02

1 a

Number of tonnes, n	0	1	2	3	4	5	10	15	20	25
Cost of delivered soil, $\$C$	60	88	116	144	172	200	340	480	620	760

b $C = 60 + 28n$

c, d

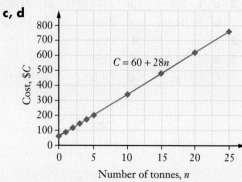

$C = 60 + 28n$

e 60

f $60, the initial fee for the delivery

g 28, the additional cost per tonne of soil in dollars

h It's the maximum amount the truck can carry.

i $1180 **j** $1240

2 a

Number of cups of chips, n	0	10	20	30	40	50	60	100	200
Cost of making chips, $C	42	49.50	57	64.50	72	79.50	87	117	192

b $C = 42 + 0.75n$ **c** $169.50

d 143

e

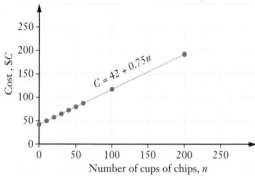

f 42, the initial cost of the oil in dollars

g 0.75, the additional cost of each cup of chips in dollars

h The graph will cross the vertical axis higher up (at 48) but it won't be as steep.

3 a 600 m **b** 5

c The graph is going down to the right, the gradient is negative, $m = -5$. The initial value is 600. $h = 600 - 5t$.

d

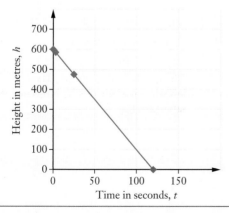

4 a

| Yearly depreciation | Salvage value, S, after n years | | | | |
	1 year	2 years	3 years	4 years	5 years
$3000	36 000	33 000	30 000	27 000	24 000
$4000	35 000	31 000	27 000	23 000	19 000

b

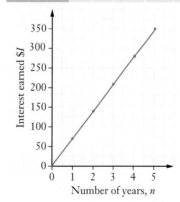

c **i** −3000, the dollar amount of depreciation each year

ii 39 000, the dollar value of the car when Eddie bought it

iii $S = -3000n + 39\,000$

d **i** −4000, the dollar amount of depreciation each year

ii 39 000, the dollar value of the car when Eddie bought it

iii $S = -4000n + 39\,000$

e 13 years **f** 7 years

5 a

Number of years, n	0	1	2	3	4	5
Interest earned, $I	0	70	140	210	280	350

b

c 70, amount of interest earned each year in dollars

d 0, amount of interest at the start in dollars

e $I = 70n$

f $490

g $2140

6 a

Litres of blood in the body of a 175 cm tall male

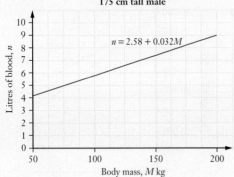

$n = 2.58 + 0.032M$

b 2.9 L

c This is an extrapolation, substituting a value outside the data range. The body mass should be between 50 and 200 kg and it's only 10 kg. Also, the male's height must be 175 cm. A 10 kg boy will not be 175 cm tall.

d 2.56 L

Exercise 3.03

1 a

x	0	1	3
y	−1	1	5

x	0	3	5
y	5	2	0

b

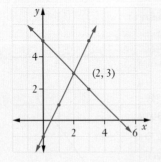

(2, 3)

c $x = 2, y = 3$

2 a

x	0	2	4
y	4	2	0

x	0	2	4
y	−5	−1	3

b

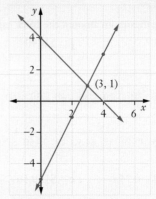

(3, 1)

c $x = 3, y = 1$

3 a $x = 2, y = 6$ **b** $x = -2, y = 3$

   c $q = -3, p = 5$ **d** $x = 3, y = -1$

   e $x = -3, y = -7$ **f** $x = 1, y = 4$

Exercise 3.04

1 a Teacher to check

b

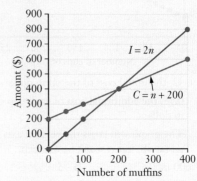

$I = 2n$

$C = n + 200$

c (200, 400)   **d** break-even point

e No, he needs to sell more than 200 to break even.

f $30

2 a Distance = speed × time

b $t = 3, D = 180$

c The number of hours after Dane left when both drivers had travelled the same distance.

3 a B   **b** 3 min   **c** B

4 a car 96 km/h, truck 72 km/h

    b 2 seconds

    c 2 seconds

    d car: $V = -24t + 96$, truck: $V = -12t + 72$

    e Both vehicles were slowing down.

    f The truck was heavier.

5 a

Bunches of flowers sold	0	10	20	30	40	50	60
Income received, $	0	100	200	300	400	500	600

    b

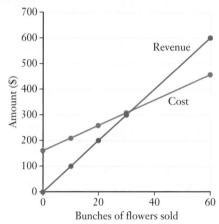

    c The graphs intersect at (32, 320), which is the break-even point.

    d More than 32 bunches

6 a income   **b** expenses

    c

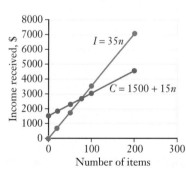

    d 75   **e** $500

7 a $60   **b** 10   **c** $6

    d It's the weekly cost of running the dog sleds, such as feeding the dogs. It doesn't change with the number of dog sled rides.

8 a $C = 10n + 160$ **b** $I = 20n$

    c

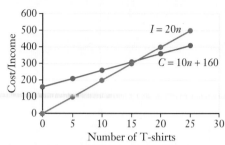

    d 16   **e** $740

9 a $40   **b** $10

    c Monthly service charge

    d 20c   **e** $D = 0.2n + 10$

    f 150   **g** plan A

Keyword activity

1	linear	**2**	straight
3	gradient	**4**	y-intercept
5	model	**6**	costs
7	interest	**8**	depreciation
9	intersecting	**10**	simultaneous
11	income	**12**	costs
13	break-even	**14**	profit
15	loss		

Test yourself 3

1 a

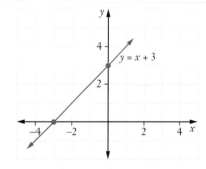

gradient = 1, y-intercept = 3

b

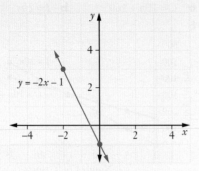

gradient = –2, y-intercept = –1

2 a i 2   **ii** –3   **iii** $y = 2x - 3$
  b i –1   **ii** 4   **iii** $y = -x + 4$

3 a

Number of people, n	0	50	100	150	200	250
Charge, $C	80	580	1080	1580	2080	2580

b

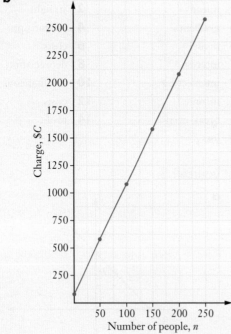

c $C = 10n + 80$
d $1780
e 75
f 80, the fixed charge in dollars
g 10, the cost per person in dollars
4 a $x = 7, y = 11$
  b $x = -1, y = 3$

5 a $C = 5n + 126$
  b $I = 12n$
  c

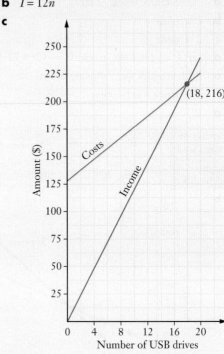

  d 18   **e** $364

Chapter 4

Exercise 4.01

1 a $33 000   **b** $22 000
2 a $1750   **b** $650
3 $24 500
4 a $3840   **b** $15 360   **c** $16 640
5 a $1300   **b** $520   **c** $2340
6 a $4815   **b** 25%   **c** 4 years
7 $1115
8 a $10 000   **b** 5 years

Exercise 4.02

1 $1119
2 a $320   **b** $305
3 a $8991   **b** $13 009
4 a $18 000   **b** $2700
  c $1951   **d** 15%
5 a End 3rd = $3200, End 4th = $2100
  b $1100   **c** $1890   **d** $210

6 a In the formula $S = 925, r = 25\%, n = 5$

b $3900

7 a Teacher to check   **b** 25%

8 a $1200

b

Year	1st	2nd	3rd	4th	5th
Depreciation	$2400	$1440	$864	$518	$311

c Declining-balance

d Straight-line

9 Increasing the rate of depreciation decreases the salvage value more quickly.

Exercise 4.03

1 a $7000 \times 0.09 \div 12 = \$52.50$

b $7000 + \$52.50 = \$7052.50$

c $7052.50 - \$320 = \$6732.50$

d

Chantelle's reducing balance car loan				
Amount borrowed: $7000				
Interest rate: 9% p.a. monthly reducing				
Monthly repayments: $320				
Month	Principal (P)	Interest (I)	Principal + interest (P + I)	Amount owing (P + I − R)
1st	$7000	$52.50	$7052.50	$6752.50
2nd	$6732.50	$50.49	$6782.99	$6462.99
3rd	$6462.99	$48.47	$6511.46	$6191.46

e $6191.46   **f** $151.46   **g** $6.06

h The interest is calculated on the amount she owes, which is getting smaller each month.

2 a

Bianca's reducing balance personal loan				
Amount borrowed: $16 000				
Interest rate: 7.2% p.a. monthly reducing				
Monthly repayments: $400				
Month	Principal (P)	Interest (I)	Principal + interest (P + I)	Amount owing (P + I − R)
1st	$16 000	$96	$16 096	$15 696
2nd	$15 696	$94.18	$15 790.18	$15 390.18
3rd	$15 390.18	$92.34	$15 482.52	$15 082.52
4th	$15 082.52	$90.50	$15 173.02	$14 773.02

b $14 773.02   **c** 22.6%

3 a $720

b   **i** $15 286   **ii** $611.44

    iii $12 463.44   **iv** $498.54

    v $6734.10

c 2 years

4 a

Jackson's loan				
Amount borrowed: $12 800				
Interest rate: 7.56% p.a. monthly reducing				
Monthly repayments: $900				
Month	Principal (P)	Interest (I)	Principal + interest (P + I)	Amount owing (P + I − R)
1st	$12 800	$80.84	$12 880.64	$11 980.64
2nd	$11 980.64	$75.48	$12 056.12	$11 156.12
3rd	$11 156.12	$70.28	$11 226.40	$10 326.40

b $2700   **c** $2473.60   **d** $226.40

Exercise 4.04

1 a $29.17   **b** $27.30

c $1024.05   **d** 15 months

e $335.93   **f** $235.93

g $5235.93

2 a 18 months   **b** $195.43   **c** $195.43

3 a 18 months   **b** $532.31

4 a 25 months   **b** 18 months

c Yes, he will save $202.26. If his repayments are $400, the interest is $718.99. If his repayments are $550, the interest is only $516.73.

d The quicker you can repay the loan, the less interest you will pay.

e Once a flat-rate loan has begun, making additional payments doesn't decrease the interest you have to pay.

5 B9=B4; C9=B7*B4/365; D9=B9+C9; E9=D7-B5; B10=E9

6 The diagram shows the final lines of the spreadsheet.

◇	A	B	C	D	E
37	31st month	$ 4,925.75	$ 30.79	$ 4,956.54	$ 4,006.54
38	32nd month	$ 4,006.54	$ 25.04	$ 4,031.58	$ 3,081.58
39	33rd month	$ 3,081.58	$ 19.26	$ 3,100.84	$ 2,150.84
40	34th month	$ 2,150.84	$ 13.44	$ 2,164.28	$ 1,214.28
41	35th month	$ 1,214.28	$ 7.59	$ 1,221.87	$ 271.87
42	36th month	$ 271.87	$ 1.70	$ 273.57	-$ 676.43
43					
44					
45		Sum of Interest = SUM(C7:C42)			
46		$	3,523.57		
47					

The spreadsheet shows that it will take 36 months for Gabriel to repay the loan. The interest is in column C. The formula =SUM(C7:C42) will produce the sum of the interest.

Gabriel will make 36 repayments and pay a total of $3523.57 in interest.

7 a 13 months   **b** $214.45

Exercise 4.05

1 a $317   **b** $19 020   **c** $4020

2 a $616   **b** $70 880

   c • The car may be worn out before Jamie has finished paying for it. He may need a new car before he has paid for the old car.

      • He will have to pay too much interest.

   d Buy a cheaper car and repay the loan as quickly as he can afford to.

3 a 17 repayments   **b** $1050

4 a $190.40

   b She owes a smaller amount and reducible interest is calculated on the amount owing.

   c $165 096   **d** $229 405

   e $919.20. It's the first month when more comes off the balance than is paid in interest.

   f 269 months; i.e. 22 years 5 months

   g 7 years 7 months

   h We pay most of the interest at the beginning of the loan.

5 Karl has significantly underestimated the amount of interest that borrowers have to pay at the beginning of a reducible interest loan.

6 a 240 months of $830 or 187 months of $930.

   b $25 290

   c They can't afford the extra $100 per month.

7 a $2301   **b** $36.40

8 a $1951.60   **b** $22.10   **c** $265.20

9 a $244.80   **b** $152 640

   c Teacher to discuss

10 $1152/month

11 a $998.20, $989.80

   b

	Big 4 bank	Small mortgage company
Loan establishment fee	$320	$598
Mortgage discharge fee	$228	$314
Total annual loan fee over 25 years	$6200	$1900
Total monthly repayments	$299 460	$296 940
Total cost of the loan	$306 208	$299 752

   c Interest rates and annual fees.

   d Go with the small mortgage company because it's cheaper by more than $6000.

Exercise 4.06

1 a $16.29   **b** 9 Nov 2018   **c** $9

   d 19.99% p.a.   **e** 04389804   **f** $27 500

2 a $993.71   **b** $1121.39   **c** $2280

   d 28 Nov 2018, $19.88

   e   **i** 8 years 11 months   **ii** $1134

     iii $2127.71

   f $2152.32

3 a $219.95, The Athletes Foot, Tuggerah

   b Dr Peter Solomon, $270

   c $164.83   **d** $923.09

   e Indulge Hair Design, $118

   f $1110   **g** 30 817

   h Two cards, 1817 points left

   i 21 750

Exercise 4.07

1 $12.69

2 $42.80

3 a 20.44%   **b** 0.056 ÷ 100 = 0.000 56

   c $57.10

4 $771.98

5 a $59.25   **b** $72   **c** $3587.25

6 a $2500

  b $2021.62, as she owes $478.38

  c $15   **d** $478.38

  e 27 April 2019

  f i $7.18   **ii** 47c

Exercise 4.08

1 17%

2 $174

3 a No, he will only have $7800.

  b If the entire payment isn't received by the due date, then an interest charge of 29.45% will be added to the amount owing. Continuing finance will be subject to interest at 29.45% p.a. calculated on the amount owing at the end of each month.

  c $7796.33

  d Approximately $48 700 over 42 years and 5 months.

4 a $82   **b** 22 years and 3 months

  c $715   **d** $556

5 a temporary   **b** immediately   **c** credit card

  d difficulties   **e** stop   **f** low

  g limited   **h** owe   **i** electricity

  j instalments   **k** every day   **l** minimum

  m huge   **n** rush

6 Discount store; it has the higher interest rate.

Keyword activity

1 depreciation   **2** salvage

3 declining-balance   **4** credit union

5 interest   **6** reducible

7 flat   **8** same

9 down   **10** real estate

11 huge   **12** repay

13 minimum   **14** fortnightly

15 money

Test yourself 4

1 a $428   **b** $92

2 a $910   **b** $710

3 a $46 269   **b** $78 731

4 a $600   **b** $120 600

  c $119 600

  d $600 interest, $400 off the loan

  e $598   **f** $183 000

5 a $109 700   **b** 40.5 months   **c** $41 909

6 a $450 960

  b 169 months; i.e. 14 years 1 month

  c $70 710

7 a 149 months = 12 years 5 months

  b $5345

Chapter 5

Exercise 5.01

1 a

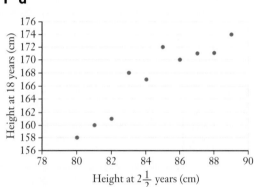

  b Height at $2\frac{1}{2}$ years is the independent variable. Height at 18 years is the dependent variable.

  c As height at $2\frac{1}{2}$ years increases, the height at 18 years increases.

2 a

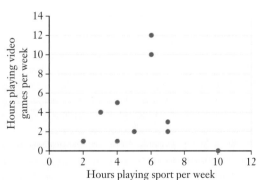

b 'Hours playing sport per week' is the independent variable and 'Hours playing video games per week' is the dependent variable.

c There is no pattern to this data. It is mostly in a bunch.

3 a

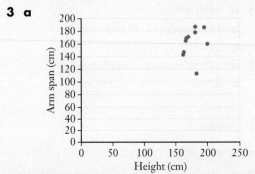

b 'Height' is the independent variable and 'Arm span' is the dependent variable.

c There is one point out on its own. The other points show that as height increases, the arm span increases.

4 a

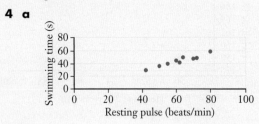

b As one variable increases, the other increases.

5 a

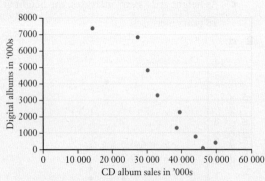

b As one variable increases, the other decreases.

Exercise 5.02

1 a positive, linear, moderate
  b no association
  c positive, linear, strong
  d negative, non-linear, strong
  e positive, linear, weak
  f negative, linear, strong
  g positive, non-linear, strong
  h no association
  i negative, linear, moderate
  j positive, linear, moderate

2 a positive, linear, strong
  b positive, linear, strong
  c negative, non-linear, moderate

3 a

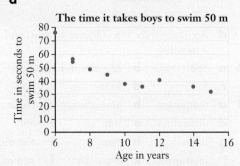

b Yes, there is a strong, negative, linear correlation.

4 Teacher to check
5 Teacher to check
6 Teacher to check

Exercise 5.03

1 a

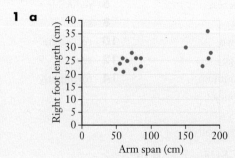

b Arm span is the independent variable, right foot length is the dependent variable.

c As arm span increases, right foot length increases.

d Positive, linear, moderate

ISBN 9780170413534

2 a No, teacher to check reasons.

b

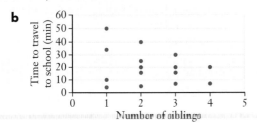

c The scatterplot shows no relationship between the two variables.

3 Teacher to check

4 Teacher to check

Exercise 5.04

1 Teacher to check

2 a, b

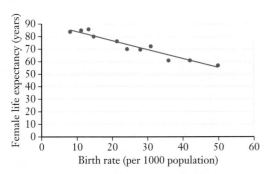

3 a, b

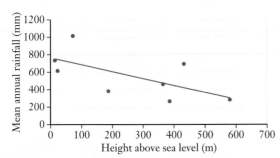

4 a, b

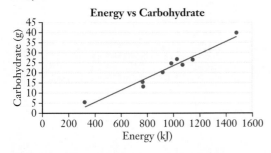

c, d

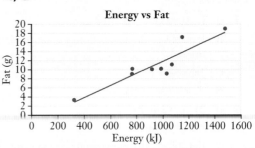

Exercise 5.05

1 a Weeks after starting training

b Leg extensions in 30 seconds

c The longer Anita has trained, the more leg extensions she can do in 30 seconds.

d 11 or 12

2 a Number of victims

b Number of medical staff

c About 40 staff

d About 750 victims

e The greater the number of victims, the greater the number of medical staff sent.

f The agency calculated the number of medical staff required based on the number of victims, not the reverse, as Kyle's statement implies.

3 a About 23.5 cm

b The longer the right foot, the higher the test score.

c The age of the students is relevant. Younger students who haven't learned as much maths generally have smaller feet and would score lower on the test.

4 a As the height of students increases, so does the weight of students.

b About 55 kg

c No, teacher to check reasoning.

5 a About 10 or 11

b As the number of sunny days increases, the number of dry cloudy days decreases.

c The total number of sunny, dry cloudy and wet days must be the same as the number of days in the month. As the number of sunny days increases, the number of dry cloudy or wet days must decrease to keep the total the same as the number of days in the month.

Exercise 5.06

1 a i 1090 kJ **ii** 105 kg

b Reliable, as the association is strong and linear.

c 1885 kJ

d Teacher to check

2 a About 62 years

b About 36 births

c Reliable, as the association is strong and linear.

d About 48 years

e Teacher to check

3 a About 525 mm

b About 460 m

c Not very reliable, as the association is weak.

d About 140 mm

e −14 mm, not possible as you can't have negative rainfall.

4 a

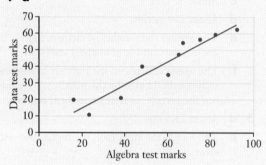

b Ella 43 in Algebra test, Amy 57 in Data test.

c Very reliable, as the association is strong and linear.

5 a

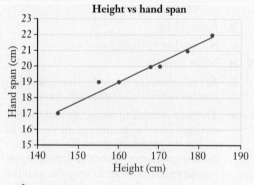

b 20 cm

c 33 cm. The value is very likely to be wrong because Robert Wadlow's height is much greater than the heights used in the scatterplot.

Keyword activity

1 bivariate

2 scatterplot

3 independent variable

4 dependent variable

5 association

6 positive, negative

7 linear, non-linear

8 strong, moderate, weak

9 line of best fit

10 predictions

11 interpolation

12 extrapolation

Test yourself 5

1 a

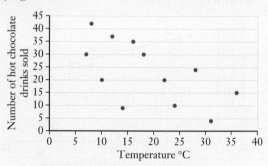

b Independent variable is 'Temperature', dependent variable is 'Number of hot chocolate drinks sold'.

c Teacher to check

2 The association is negative, linear and moderate.

3 Teacher to check

4 a Teacher to check

b

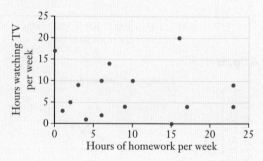

c Independent variable is 'Hours of homework per week', dependent variable is 'Hours watching TV per week'.

d There is no association between these two variables.

5

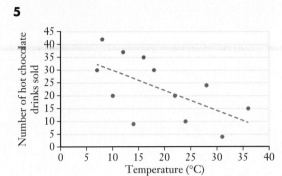

6 a 22 hot chocolate drinks

b 16°C

7 a, b

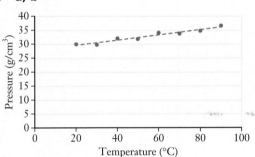

c 33 g/cm³

d Very reliable, as association is linear and strong.

e 26.8 g/cm³

f Not very reliable, as this is a long way outside the given data. The association may not continue to be linear and we don't know what will happen to the pressure as it gets colder.

Practice set 1

Section 1

1 D		**2** C		**3** C
4 A		**5** C		**6** D
7 B		**8** D		**9** B
10 A		**11** B		**12** A

Section 2

1 a $6150 **b** $5250

c Approximately 6 years

2 75 white tiles

3 a $4156.77

b $656.77

4 568 students

5 a

Number of sausage sandwiches, n	0	10	20	30	40	50	100	150	200
Cost, C	32	41	50	59	68	77	122	167	212

b $C = 32 + 0.9n$

c $185

d

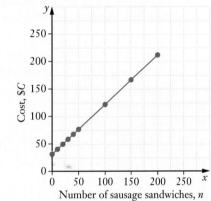

e 32, initial cost of gas bottle in dollars

f 0.9, represents the cost per sandwich in dollars

6 a

Number of sausage sandwiches, n	0	10	20	30	40	50	100	150	200
Income, I	0	20	40	60	80	100	200	300	400

b $I = 2n$

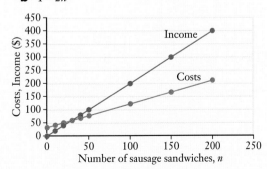

c Approximately (30, 60)

d Break-even point at which Year 12 doesn't lose or make money.

e Yes, the income line is above the costs line.

f $188

7 $2560

8 a

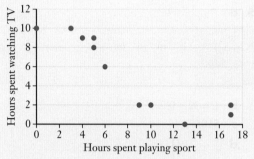

b As the hours spent playing sport increases, the hours spent watching TV decreases.

c The association is negative, linear and moderate.

10 a Approximately 960 m

b Approximately 160 m

c i 7.5 cm   **ii** Yes

11 a

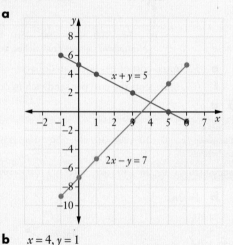

b $x = 4, y = 1$

12 a $8707.60   **b** $307.60

13

Rose and Ian's reducible finance loan				
Amount borrowed: $18 000				
Interest rate: 8.4% p.a. monthly reducing				
Monthly repayments: $475				
Month	Principal (P)	Interest (I)	Principal + interest (P + I)	Amount owing (P + I − R)
1st	$18 000	$126	$18 126	$17 651
2nd	$17 651	$123.56	$17 774.56	$17 299.56
3rd	$17 299.56	$121.10	$17 420.66	$16 945.66
4th	$16 945.66	$118.62	$17 064.28	$16 589.28

14 a i 209 m   **ii** 475 m

     b i 3.1 cm   **ii** 14.7 cm

15 $80 217

16 a $78.63   **b** $94.40   **c** $4704.23

19 a

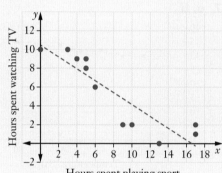

b It is negative, linear and moderate.

c Approximately 5.5 hours of watching TV

d Approximately 12 hours playing sport

e Not very reliable, as the association is only moderate, and not very strong.

Chapter 6

Exercise 6.01

1

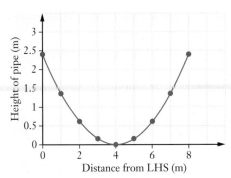

2 a

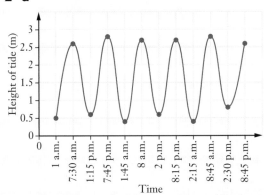

b Approximately 0.9 m

c Approximately 4.30 p.m.

3 a Time

b

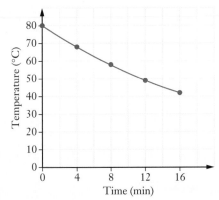

c Approximately 74°C

d Approximately 10 minutes

4 a

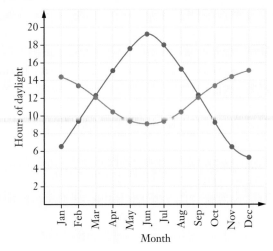

b They both begin and end at about the same value for hours. Other answers are possible.

c Anchorage starts low, goes up and comes down again. Perth is the opposite. Other answers are possible.

d March, September

e Anchorage is further north of the Equator than Perth.

5 a

Number of people	1	5	10	20	30	40
Cost per person ($)	840	168	84	42	28	21

b

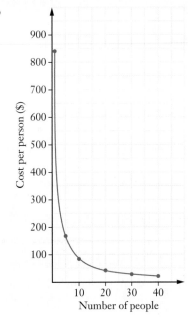

c Approximately 30

d $28

Exercise 6.02

1 a 800 m **b** 600 m **c** 200 m **d** 360 m

2 5.5 m

3 a 45 m **b** 12 m **c** 100 m

4 a 33 m **b** 80 m **c** 68 m

5 a

Horizontal distance, d m	−32	−24	−16	−8	0	8	16	24	32
Height, h m	51.2	28.8	12.8	3.2	0	3.2	12.8	28.8	51.2

b

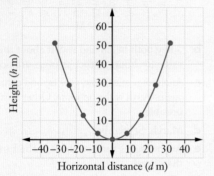

c 51.2 m

d Approximately 20 m and −20 m

6 a

Horizontal distance, d m	0	1	2	3	4	5	6	7	8
Height of curve, h m	0	7	12	15	16	15	12	7	0

b

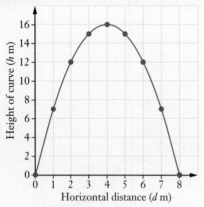

c 16 m **d** Approximately 1.2 m and 6.8 m

7 a Perimeter $= x + (18 − x) + x + (18 − x) = 36$ cm

b 10 cm × 8 cm

c 80 cm^2

d Negative lengths and widths are impossible.

e $A = x(18 − x) = 18x − x^2$

f

Length of side, x cm	0	2	4	5	7	9	11	13	15	17	18
Area of rectangle, A cm^2	0	32	56	65	77	81	77	65	45	17	0

g

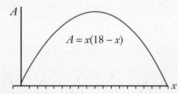

h 81 cm^2

i square

Exercise 6.03

1 a

Year	0	1	2	3	4	5	6
Value of investment ($)	800	872	950	1036	1129	1231	1342

b

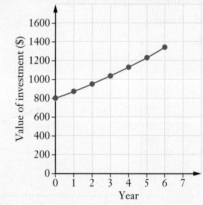

2 a $2252.19

b

Year	0	2	4	6	8	10	12
Value of investment ($)	1000	1145	1311	1501	1718	1967	2252

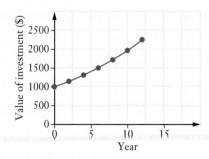

c Approximately $1450

d Approximately 7 years

3 a Teacher to check

b

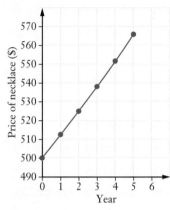

c Approximately 3.5 years

d $38.45

4 a 2 years should be $32 448, 4 years should be $35 096.

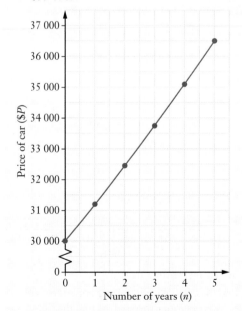

b $33 746

c Teacher to check

d It is unlikely to be 4% every year. Other factors are economic conditions, marketing and popularity of the car. Other answers are possible.

5 a Teacher to check

b

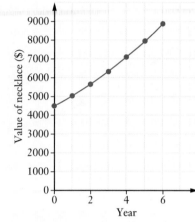

c Approximately $5900

6 a Teacher to check

b

Year (n)	0	1	2	3	4	5	6
Salvage value ($$S$)	30 000	25 500	21 675	18 424	15 660	13 311	11 314

c

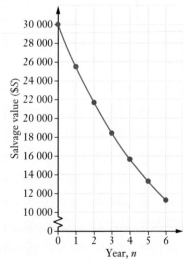

d Approximately 4 years 3 months

7 a $V_0 = 7700$ and $r = 0.2$

b

Year (n)	0	1	2	3	4	5
Salvage value (S)	7700	6160	4928	3942	3154	2523

c

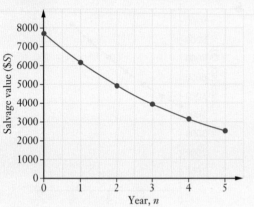

d A little over 3 years

e After approximately 5 years

Exercise 6.04

1 a 15 rabbits

b 75 rabbits

c 94 rabbits

d

Date	1 Jan	1 Feb	1 March	1 April	1 May	1 June
Number of rabbits	60	75	94	117	146	183

e

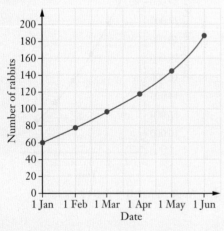

f Slightly more than 3 months (about 3 April).

g 12 705 rabbits

h It is unlikely. Other factors include disease, food supply, hunting.

2 a

Year BCE	7000	6000	5000	4000	3000	2000	1000	0
Population (millions)	2.4	4.8	9.6	19.2	38.4	76.8	153.6	307.2

b

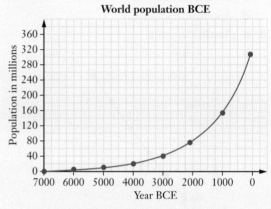

c No, because if it had been, the population in 2000 would have been only 1228.8 million.

3 a

Number of days (t)	0	1	2	3	4	5	6	7	8	9	10
Number of infected people (I)	40	42	45	48	50	54	57	60	64	68	72

b

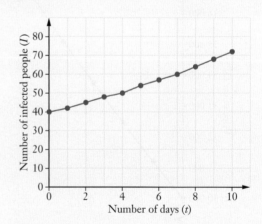

c A little over 5 days

d No. People will recover and others will take precautions to avoid getting the flu.

4 a 12

b 20%

c The percentage increase is very close to the same each year.

d

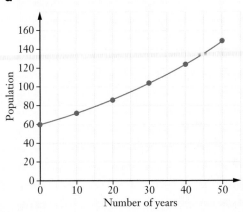

e 372

5 a

Day, t	0	1	2	5	7	10	12
Weight, W (g)	13	14	15	18	21	26	29

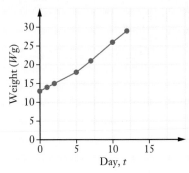

b A little over 10 days

c No. The growth will slow as the kangaroo gets bigger.

6 a

Years	0	1	2	3	4	5
People	50	100	200	400	800	1600

b

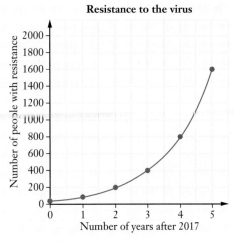

Resistance to the virus

c 52 428 800

Exercise 6.05

1 a C **b** A **c** B

2

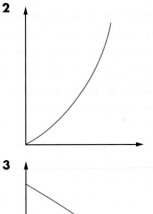

3

4–7 Teacher to check

8 a Teacher to check

b

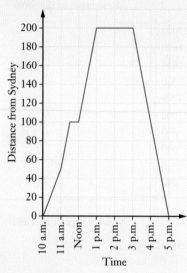

9 Teacher to check

Test yourself 6

1

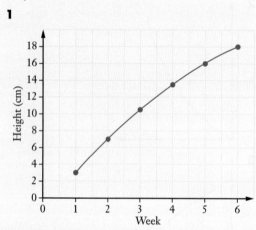

2 a Approximately 48 m

b 80 km/h

c 43 m

3 a

Horizontal distance from centre of road, x (m)	−50	−40	−30	−20	−10	0	10	20	30	40	50
Height above road, h (m)	0	18	32	42	48	50	48	42	32	18	0

b

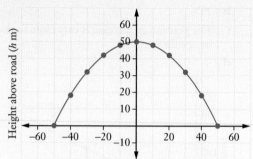

c 50 m

d Approximately −35 m and 35 m

4 a

Year	0	1	2	3	4	5
Value of investment ($)	9000	9360	9734	10 124	10 529	10 950

b

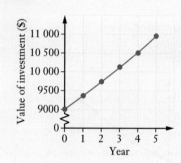

c Approximately 1 year and 6 months

5 a $V_0 = 47\,000, r = 0.15$

b

Year	0	1	2	3	4	5	6
Salvage value ($)	47 000	39 950	33 958	28 864	24 534	20 854	17 726

c

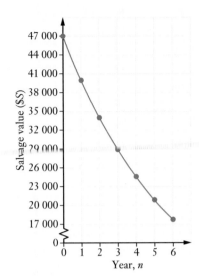

d A little over 4 years

6 a

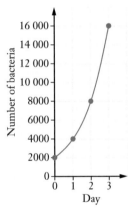

b Approximately 1 year and 3 months

c 5 120 000

7 Teacher to check

8 Teacher to check

Chapter 7

Exercise 7.01

1 a 50 km/h   **b** 25 words/min

   c 50 L/h   **d** $17.50/kg

   e 7.27 m/s   **f** $26/h

   g 250 g/L   **h** 600 revs/min

2 18 L/m^2

3 1350 L/h

4 a 40 L/container   **b** 4 L/day

   c $5\frac{1}{3}$ g/cm^3 **d** $3.20/DVD

   e 35 mm/day   **f** 14.5 km/L

   g $11.40/m   **h** 256 vibrations/s

   i 40 sheep/ha   **j** $1.58/L

5 9 m^2/L

6 $120/day

7 a $8.50/kg   **b** $12.48/kg

   c $2.45/kg   **d** 55c/roll

   e $2.45/bottle

Exercise 7.02

1 a $T = 5$ **b** $D = 240$

2 a 180 km   **b** 4 h

3 80 km/h

4 275 km

5 8 km

6 a 0.15 h   **b** 9 min

7 7.5 min

8 a 2000 m = 2 km

   b 30 is minutes not hours

   c 4 km/h

9 a 279 m   **b** 4.5 s

10 a 5400 m   **b** $11\frac{1}{9}$ s

11 a 750 km   **b** 10 h

   c 76 km/h   **d** 5 a.m. Tuesday

12 a 8 km   **b** 4 km

   c 1 hour   **d** 4 km/h

   e 2 hours   **f** 6 km/h

   g The second hour is uphill. In the first hour his speed was 8 km/h and in the second hour his speed was only 4 km/h.

13 a i 2 p.m.   **ii** 11 a.m.

   b 12:45 p.m.   **c** 48 km   **d** 4.5 h

   e i 16 km/h   **ii** 6.4 km/h

     iii 9.6 km/h   **iv** 9.6 km/h

14 a 240 km

   b Sam, steeper graph, ending journey earlier.

   c 120 km   **d** 180 km

   e 10 a.m., 160 km

   f after 9:15 a.m., graph is steeper

   g 40 km/h   **h** B

15

Graph	Person	Distance from school
A	Luke	10 km
B	Peta	8 km
C	Walter	4 km
D	Shelby	2 km

16 Teacher to check

Exercise 7.03

1 a 5.6 m/s **b** 13.9 m/s
 c 22.2 m/s **d** 30.6 m/s
2 a 4.8 km/h **b** 1.3 m/s
3 a 45 km/h **b** 144 km/h **c** 3.5 km/h
4 40 kg/ha
5 1.8 L/h
6 a 1.2 g/mm **b** 1.2 t/km
 c the same, in both cases × 100 and ÷ 1000
 d 12 g/cm
7 a 3500 L/h **b** 58.3 L/min **c** 1.0 L/s
8 a $2.80/L **b** 0.28c/mL
9 a 10.4 m/s **b** 37.44 km/h

Exercise 7.04

1 a $2.30 **b** $2.20 **c** $2.25
2 a 50 mL: 72c/10 mL; 80 mL: 70c/10 mL
 b 80 mL
3 a 40c and 35c
 b box containing 6 eggs
 c 3 boxes containing 6 eggs
4 350 g, as it's the cheapest per 100 g
5 a 1 kg **b** 1 kg for $3.50
6 a $145, $140, $139, $142
 b one 2.5 g packet and one 5 g packet
7 Teacher to discuss
8 The 1 L bottle is the better value. The 750 mL bottle is the equivalent of $11.93 per L.

Exercise 7.05

1 a 112 **b** 196 **c** 336
2 $248

3 a 3 h **b** 1.5 h
 c 4.8 h or 4 h 48 min
4 6.25 kg
5 a 720 kg **b** 15 trees
6 a 5460 km
 b **i** 2.0 h **ii** 3.1 h
   iii 4.1 h **iv** 15.2 h
7 a 960 **b** 270
 c 25 m^2 **d** 95 m^2
8 768 kg
9 a 750 mL **b** 1500 mL **c** 1.5 L
 d 36 L **e** 40 min **f** 40 h
10 a $8000 **b** 35 weeks **c** 3 weeks
11 a 18 h **b** 35 h **c** 24 h
12 a 21.6 m **b** $270 **c** 18 m
 d 10 windows
13 a 504 L **b** $1134
 c $2232 **d** 15 h

Exercise 7.06

1 a 50.15 L **b** $80.24
2 13.9 L/100 km
3 820 km
4 9.5 L/100 km
5 a 10.1 L/100 km **b** 8.8 L/100 km
 c 8.5 L/100 km **d** 7.3 L/100 km
 e 7.8 L/100 km
6 800 km
7 a 7.15 L **b** $10.73
8 a 47 L **b** $82.25
9 a 0.4 L/h **b** 3.2 L
 c 7.5 h
10 a They both use 1.8 L to complete the job.
 b Use whichever mower he wants. However, with the ride-on, there will be less noise pollution.

Keyword activity

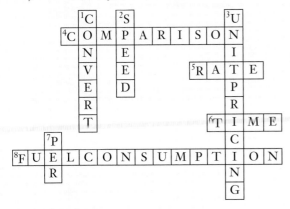

Test yourself 7

1 $64/h

2 a 240 km   **b** 20 km

3 320 km/h

4 20.8 m/s

5 108 000 km/h

6 a Small $1.33   Medium $1.07   Large $1.30

   b Medium, 750 mL

7 $2310

8 $106

9 Choose $36.50/h. It pays $60 more per week.

10 a White car 8.5 L/100 km, Red car 9.1 L/100 km

    b White car

Chapter 8

Exercise 8.01

1 a $V = 4, E = 3$ **b** $V = 4, E = 2$

   c $V = 7, E = 12$ **d** $V = 4, E = 4$

   e $V = 4, E = 4$ **f** $V = 6, E = 5$

   g $V = 5, E = 7$ **h** $V = 2, E = 3$

   i $V = 5, E = 7$

2 a $ABCD, AFD, AED$

   b 27 km, 25 km, 24 km

   c AED

   d $A\,3, B\,2, C\,2, D\,3, E\,2, F\,2$

   e Yes, provided we start and finish at A or C, the two odd vertices.

3 a A to B via the curve is 17 km, ADB 16 km, ACB 17 km.

   b 15 km

c $ABCD$

d A, B, C, D

4 a $ADC, ABC, ABDC, ADBC$

   b $ADBC$

   c Start at B or D and finish at the other.

5 a $AEDC$ 17 km, $AEBDC$ 16 km, $ABEDC$ 24 km, $ABDC$ 17 km; $AEBDC$ is the shortest route.

   b E, B, D, C

   c No

6 a 9 km

   b E

   c There are more than two odd vertices in the network. If you try to traverse the network, you get stuck on one of the odd vertices and can't get out without going over a previously travelled road.

7 a There are exactly two odd vertices.

   b At either the winery or the restaurant

8 The shortest routes involve going directly to South side, then taking the appropriate road to the fire.

Exercise 8.02

1 a $B9, D3$

   b It's only 8 km to get to B via D.

   c 14 km

   d $ADBC$

2 a $B6, D10$

   b Cross out 10, write $D9$; i.e. $B6\ D9$

   c 16

   d $ABDC$ via the curved route, 16 km

3 19 km, $ACDEF$

4 a $ACED$ **b** 1:30 p.m.

5 a $ABCD$ **b** 3 hours 7 minutes

6 3 km

7 Teacher to check

Exercise 8.03

1 a

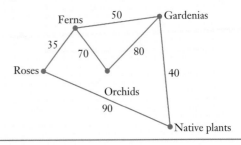

b   85 m (via the ferns)

c   Gardenias→ orchids

Gardenias→ ferns→ orchids

Roses→ ferns→ orchids

d   120 m

2 a

	Bus stop	Jane's house	Part-time job	Pool	School
Jane's house	5				
Part-time job	–	–			
Pool	–	–	–		
School	7	–	6	–	
Shops	–	10	12	9	–

b   12 minutes

c   Many answers possible. There may be no footpath to walk on. Jane might want to walk via the shops. She might be planning to meet friends.

3 a   Choa Chu Kang

b   Follow the red line to City Hall where she should change to the green line, which will take her directly to the airport.

4 a   3.7 km via the BBQ or 3165 m via the wildflower garden.

b   Approximately 1 hour

Exercise 8.04

1 a   Lady beetles

b   Lady beetles, Caterpillars, Beetles

c   There would be less food variety for Butcherbirds and nothing to eat the aphids. The aphids could reach plague proportions.

2 a iii   **b ii**   **c iv**   **d i**

3 a

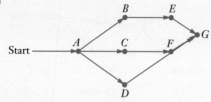

b

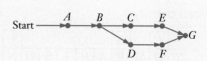

c

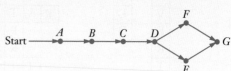

d

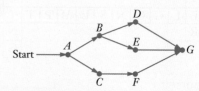

4   One possible answer

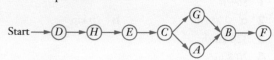

5   *A* **x**   *B* **ii**   *C* **iii**   *D* **xii**

E **vii**   *F* **ix**   *G* **xi**   *H* **v**

J **viii**   *K* **iv**   *L* **vi**

Exercise 8.05

1 a   It's two separate pieces.

b   There's a closed circuit.

c   One vertex isn't connected.

d   There's a closed circuit and 3 vertices aren't connected.

2 a   Teacher to check

b   *FE* isn't an edge in the original network.

3 a

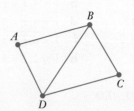

b   Teacher to check

4

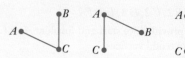

5 a $4^{(4-2)} = 4^2 = 16$

b   Teacher to check

1 a

Length = 21

b

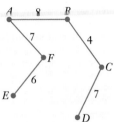

Length = 32

c

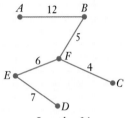

Length = 34

d

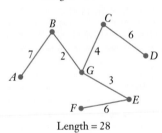

Length = 28

2 a

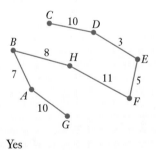

b Yes

3 a

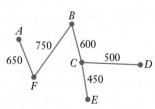

b $2950

4 a

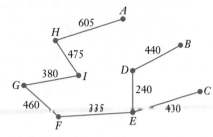

b 2885 m

5 Minimum amount of wire needed is 13.5 mm.

Exercise 8.07

1 a

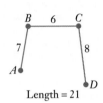

Length = 21

b

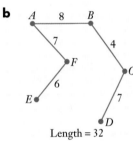

Length = 32

c

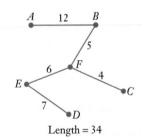

Length = 34

d

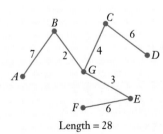

Length = 28

2 a

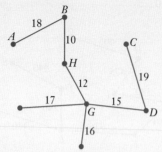

b 107 m

3 a

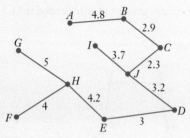

b 33.1 m

4 a

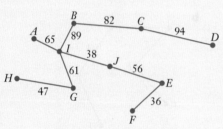

Minimum length = 568 m

4 b

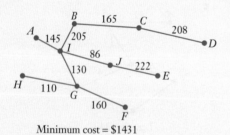

Minimum cost = $1431

c Teacher to check

Test yourself 8

1 a 6   **b** 9   **c** 2
  d 4   **e** C, F
  f Yes, start/finish at C and F.

2 ACDEF, 15 km

3 a

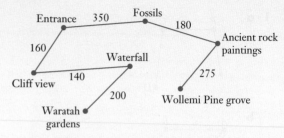

b David will have to walk via the Fossils and the Ancient rock paintings. The walk is 805 m long.

4 a I   **b** B

c

Task	Prerequisite
A	Start
B	A
C	B
D	B
E	B
F	C
G	D, F
H	G, E
I	H

5 a Vertex F isn't connected.
  b The edge CG isn't part of the original network.
  c The two sections aren't joined.
  d There's a closed circuit.

6 Teacher to check

7

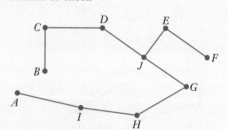

8 Minimum length of pipe required is 171 m.

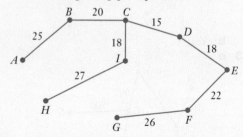

Chapter 9

Exercise 9.01

1 8.4 m   **2** 23.8 m
3 5.831 m   **4** 152 km/h
5 a yes   **b** yes   **c** no   **d** yes
6 12.37 m   **7** 733 m
8 44.4 m   **9** 148 km
10 Teacher to check

Exercise 9.02

1 a   **b**

c

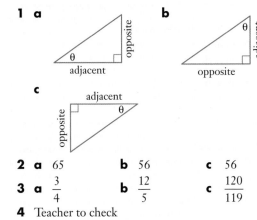

2 a 65   **b** 56   **c** 56
3 a $\dfrac{3}{4}$ **b** $\dfrac{12}{5}$ **c** $\dfrac{120}{119}$
4 Teacher to check
5 40°; 9, 40°, 9; 7.6
6 a 14.43 cm   **b** 4.86 m   **c** 10.89 cm
    d 235.12 cm   **e** 6.64 km   **f** 4.71 mm
7 a 17.16 cm   **b** 12.0 cm

Exercise 9.03

1 a 40° 42'   **b** 46° 56'
    c 36° 52'   **d** 30° 58'
    e 53° 58'   **f** 53° 8'
2 a 37° 52'   **b** 33° 41'
    c 68° 12'   **d** 48° 49'
    e 42° 31'   **f** 59° 2'
3 33° 41' = 34° and 56° 19' = 56°
4 a 45°
    b When the numerator (top) of a fraction and the denominator (bottom) are the same, the fraction is 1.

Exercise 9.04

1 a 5.8   **b** 12.6   **c** 11.4
2 a 39° 3'   **b** 28° 9'   **c** 26° 23'

3 a 15 m   **b** 8 m   **c** 17 m
4 a 41°   **b** 62°   **c** 63°
5 a 5.5   **b** 11.1   **c** 3.5

Exercise 9.05

1 a tan   **b** cos   **c** sin
2 a 7.2   **b** 6.9   **c** 10.8
3 a 63° 46'   **b** 29° 50'   **c** 52° 57'
4 a 40°   **b** 30°   **c** 40°
5 7.3 m
6 a 14.6   **b** 7.3   **c** 7.6
7 a 41°25'   **b** 29°45'   **c** 46°24'

Exercise 9.06

1 a 13.6   **b** 24.8   **c** 16.01
2 a 12.1   **b** 15.9   **c** 16.1
3 a 21.45 m   **b** 21.5 m
4 a 17.3   **b** 7.0   **c** 9.9
    d 10.1
5 a 7.4   **b** 6.0   **c** 13.5

Keyword activity

Test yourself 9

1 a 7.5 **b** 5.2

2 Yes, both contain right angles. Pythagoras' theorem works in both triangles.

3 9.0

4 a 60 **b** 60

5 a 42° 26' **b** 61° 12'

6 a 37° 58' **b** 67° 31'

7 10.6

8 68° 34'

9 a sin, $x = 3.6$ **b** cos, $\theta = 39°$

   c tan, $h = 10$ **d** tan, $\theta = 25° 44'$

10 11.4

Practice set 2

Section 1

1 C **2** B **3** B **4** C

5 A **6** A **7** D **8** C

9 D **10** D **11** A **12** B

Section 2

1 a

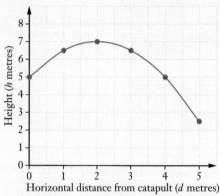

Horizontal distance from catapult (*d* metres)

b 7 m

2 3024 mL

3 a

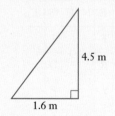

4.5 m

1.6 m

b 4.8 m

4 a

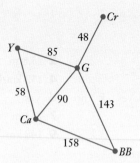

b 138 km

c Any three of Y to G to BB, Y to Ca to BB, Y to G to Ca to BB, Y to Ca to G to BB

d Y to Ca to BB, 216 km

5 7.75 cm

6 63.9 km/h

7 a Teacher to check

b

Year (*n*)	0	1	2	3	4	5	6
Salvage value ($)	14 990	12 742	10 830	9206	7825	6651	5653

c

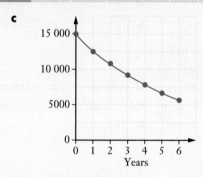

Years

d A little more than 4 years.

8 Teacher to check

9 $\theta = 43°$

10 a

Number of days (*t*)	0	1	2	3	4	5	6	7	8	9	10
Number of infected people (*l*)	30	32	34	37	39	42	45	48	52	55	59

b

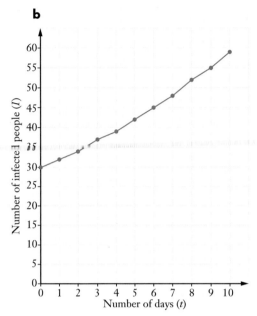

c Sometime during the 5th day.

d Teacher to check

11 a

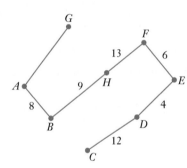

b 64 km

12 a 675 km

b $59.95

13 Teacher to check

14 50.4 L/day

15 a

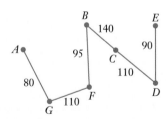

b 625 km

c Condition of roads, weather, how long each visit takes, where she can stay.

16 32.8 m

1 117/min

2 a 126 **b** 90 **c** 148

d 153 **e** 156 **f** 156

3 4920

4 a 168 **b** 332

5 a 25–38 **b** 3–4 years

c Yes, her pulse rate is only 57 beats/minute and it should be 70–110.

6 a 149 beats/min

b 44–45 days

c Teacher to check

7 Premature. His pulse is 156 beats/min.

Exercise 10.02

1 a 205 **b** 208 **c** 200

d 209 **e** 199 **f** 207

2 a 420 **b** 23 100

3 1770 **4** 6075

5 a 209 **b** 167

6 23 years **7** 11 years

8 b $(208 − 55) \times 0.7 + 55 = 162$ and $(208 − 55) \times 0.8 + 55 = 177$

c $(200 − 73) \times 0.6 + 73 = 149$ and $(200 − 73) \times 0.7 + 73 = 162$

d $(209 − 50) \times 0.8 + 50 = 177$ and $(209 − 50) \times 0.9 + 50 = 193$

e $(199 − 59) \times 0.7 + 59 = 157$ and $(199 − 59) \times 0.8 + 59 = 171$

f $(207 − 68) \times 0.6 + 68 = 151$ and $(207 − 68) \times 0.7 + 68 = 165$

9 a 4550 mL = 4.55 L

b approx. 117 mL

Exercise 10.03

1 a 72 mmHg **b** 125 mmHg

2 18 mmHg

3 a No, his diastolic pressure is in the pre-hypertension category.

b Her systolic pressure is greater than 120.

c Stage 1 high blood pressure

d Seek urgent medical advice

4 a Yes

b 3–5 years

c Approximately 10%

d It increases

Exercise 10.04

1

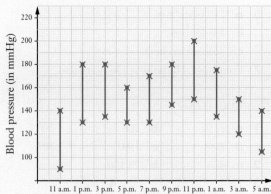

2 a 82–90   **b** 131 and 86   **c** 110–134

d 120/79   **e** Teacher to check

f Systolic range is much bigger than the diastolic range.

3 a 1:30 p.m.   **b** 170/110

4 a Blood pressure is in the normal range during the day and high when the patient is asleep.

b Nocturnal means during the night. This is high blood pressure that happens at night.

5 a 4 a.m.

b Sleeping

c 9 p.m., 220/90

d Systolic blood pressure is very high, but the diastolic pressure is normal.

e The arteries are hard and inflexible so they can't flex when the heart pumps, which creates high systolic blood pressure. When the heart isn't pumping, the arteries don't need to flex, so the blood pressure isn't affected, as shown by the normal diastolic blood pressure.

Keyword activity

Teacher to check clues

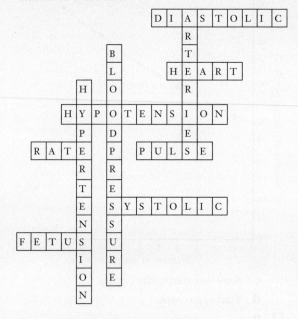

Test yourself 10

1 80 beats/minute   **2** 2460

3 209 beats/minute   **4** 16 years

5 182 beats/minute   **6** 118 mmHg

7 a A high blood pressure reading only when the doctor is taking the blood pressure.

b 6 p.m.

c 160/100

d No. He has white coat syndrome.

e Midnight to 6 a.m.

Chapter 11

Exercise 11.01

1 9.56   **2** 36° 22'   **3** 3.40

4 29° 38'   **5** 18.43   **6** 33° 41'

7 13.92   **8** 63° 38'

9 7.7 cm and 9.2 cm

Exercise 11.02

1 14 m   **2** 5° 29'   **3** 12.4 m

4 32.5 m   **5** 200 m   **6** 11.5 m

7 5.05 m   **8** 3.10 m

9 a 30°
　　b $AC = 0.69$ m, $AD = 1.38$ m
10 a 5.2 m
　　b 23.2 m
　　c 43° 30'
　　d 33.7 m

Exercise 11.03

1 118.3 m
2 1411.4 m
3 530.0 m
4 a 156.934 m
　　b 158.5 m
5 a

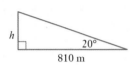

　　b 295 m
6 801.6 m
7 a 38°
　　b 52°
8 a

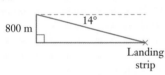

Landing strip

　　b 3209 m
9 97.5 m
10 a Teacher to check
　　b 87 cm
　　c 64° 14'

Exercise 11.04

1 a NW　　**b** S　　**c** N　　**d** SE
2 a SW　　**b** Poland, West Germany, UK
3 a NW　　**b** SW　　**c** S
　　d W　　**e** E　　**f** NE
　　g SE　　**h** N　　**i** E
4 A
5 a S 58° E　　**b** S 51° W　　**c** N 22° W
　　d N 57° E

6 a Condobolin　**b** Molong
　　c West Wyalong
7 a rain　　　　**b** hot, dry　　**c** rain
　　d cold　　　　**e** hot, dry　　**f** wet
　　g wet　　　　**h** hot, dry

Exercise 11.05

1 a 045°, N 45° E　　**b** 315°, N 45° W
　　c 270°, W　　　　**d** 225°, S 45° W
　　e 045°, N 45° E　　**f** 135°, S 45° E
2 a 110°, S 70° E　　**b** 158°, S 22° E
　　c 220°, S 40° W　　**d** 250°, S 70° W
　　e 280°, N 80° W　　**f** 014°, N 14° E
3 a 314°　**b** 272°　**c** 136°　**d** 245°
4 a 100°　　　　　　**b** S 80° E
5 a 099°　　　　　　**b** S 81° E

Exercise 11.06

1 a Teacher to check　**b** 658 m
　　c 1348 m
2 a 29°　　　　　　　**b** 138 km
3 a $\alpha = 25°, \beta = 65°$　**b** 90°
　　c 192 km
4 a Teacher to check　**b** 124 km
5 a Teacher to check　**b** 85 km
6 70 km

Exercise 11.07

1 a distances　**b** east　　　**c** challenge
　　d ocean　　　**e** dolphins　**f** compass
　　g songs　　　**h** navigator　**i** birds
　　j star　　　　**k** set　　　　**l** behind
　　m latitude　　**n** clouds　　**o** slowly
　　p New Zealand **q** White　　**r** opposite
　　s season　　　**t** nest
2 a east　　　　　　**b** west
3 They followed the main axis in the opposite
　　direction to south.
4 a 412 hours or just over 17 days
　　b 211°

Keyword activity

		C	O	S			D			
	B	E	A	R	I	N	G	E		D
O	C	O	M	P	A	S	S	P	T	E	H
P	E	A			E	Z	R	A	G	Y
P	A	D			L	E	E	N	R	P
O	S	J			E	N	S		E	O
S	T	A	S		V	I	S	S	E	T
I	N	C	L	I	N	A	T	I	O	N	E
T	E	E	N		T	H	O	U		N
E	N	E		I	S	N	T		U
	T	R	U	E	O	T		H		S
M	I	N	U	T	E	N	A	N	G	L	E
	N	O	R	T	H		R	W	E	S	T

The words included in the Keyword activity above:

sine	east	north	degree	opposite	elevation	inclination
cos	west	south	compass	adjacent	bearing	depression
tan	true	angle	minute	hypotenuse	zenith star	

Test yourself 11

1 a 4.5 **b** 4.2
  c 16.2 **d** 19.9
2 a 31° 17' **b** 43° 10' **c** 50° 19'
3 65 cm
4 38°
5 a 2570 m **b** 6.4 m
6 C
7 a C **b** B **c** A
8 576 m
9 056°

Chapter 12

Exercise 12.01

1 a 2 **b** 4
  c $\frac{2}{3}$ **d** $\frac{5}{4}$ or 1.25
2 a 6, $x = 24$ cm **b** 1.5, $x = 6$
  c $\frac{3}{4}$, $x = 22.5$ **d** $\frac{1}{3}$, $x = 3\frac{1}{3}$, $y = 2\frac{2}{3}$

3 a 0.666 ...
  b Both sides join angles marked with a dot to angles marked with an arc. $CD = 16$
  c $CE = 12$
4 a 1.25 **b** 25 **c** 5
5 a 2 **b** 7 m and 4.8 m
6 a $\frac{1}{3}$ **b** 3.6 m

Exercise 12.02

1 a 12 m **b** 30 m
  c 4.5 m **d** 2 cm
  e 16.5 m **f** 208 cm = 2.08 m
2 1100 m = 1.1 km
3 25 m
4 Length 4 m, width 2.5 m
5 a 13.5 cm **b** 7 m
  c 16 cm **d** 64 cm
6 4 m
7 6.4 km

Exercise 12.03

1 a

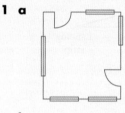

  b

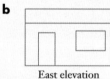

East elevation   West elevation

2–5 Teacher to check

Exercise 12.04

1 a i kitchen sink
  ii vanity (or wash basin)
  iii toilet **iv** window
  v bath **vi** built-in wardrobe
  vii sliding door
  b They are all tiled **c** 3
  d 1 and 2

e 4100 mm × 3200 mm

f in the back of the garage

g near the bathroom

h 1 **i** 1 **j** 2

k living room

l 6670 mm × 5500 mm

m i south **ii** north **iii** south
iv east **v** west

n south **o** no

p living room, dining room, bedroom 1, bedroom 3

q dining room, bedroom 2, living room

2 a 3 **b** 3 **c** no

d $86 016

3 a 5 m **b** 2.5 m **c** 2 m

d 2 m × 2.5 m = 5 m² **e** 5 m

f 4 m × 3.8 m = 15.2 m²

4 a 240 mm **b** 70 mm **c** 12.39 m

d 15.69 m **e** 3.6 m by 6.07 m

5 a 6790 mm × 3700 mm **b** 2

c outside the bathroom

d shower **e** bedroom 4

f family room = 25.123 m², media room = 23.787 m². Family room is bigger by 1.336 m²

Exercise 12.05

1 a 7.5 m² **b** $273.75

2 a 18 m **b** $198

3 a 92 m² **b** $77 280

4 a 15 m **b** 14 m **c** $210

5 a 14 m² **b** 4 m **c** $317.80

6 a 14 m²

b i $2\frac{1}{3}$ L ≈ 3 L **ii** $34.90

7 a Painting, skirting boards, carpet. Painting the ceiling needs to be done first to avoid paint splashes on the skirting boards and, most importantly, to avoid getting paint on the carpet.

b $562.70

8 $1258

9 a 3.9 m **b** 3.9 m **c** 3 m

10 $892

11 a 11.53 m **b** 3

12 Approximately $87 912

Exercise 12.06

1 a No **b** Yes

2 a 697.5 m² **b** 98.28 m² **c** 599.22 m²

3 a

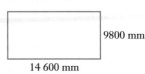

9800 mm

14 600 mm

b 143.08 m²

4 a 98.28 m² **b** 11 bags **c** $578.60

5 a 28.56 m² **b** 29% **c** 5.2 kW

d Teacher to check

6 bedroom 1

7 a 8.1 m² **b** 21.84 m² **c** 37%

8 24.57 m²

9 a 210 m² **b** Teacher to check

Exercise 12.07

1–3 Teacher to check

4 a Teacher to check **b** 2.9 km

5 a 285 m **b** 285.6 m **c** Very close.

6 a Teacher to check **b** 144.2 cm

7 133 m

8 251 m

Keyword activity

1 I **2** N **3** L **4** M

5 F **6** J **7** K **8** E

9 P **10** A **11** O **12** C

13 G **14** D **15** B **16** H

Test yourself 12

1 a i 0.25 **ii** x = 8 cm, y = 2.5 cm

b i 3 **ii** x = 3 cm, y = 6 cm

c i 0.4 **ii** k = 22.5 cm

d i 4 **ii** d = 6.25 cm

e i $\frac{1}{3}$ **ii** $p = 3\frac{1}{3}$ cm

2 a 91 cm approximately

b 11.2 cm approximately

3 a Teacher to check   **b** $x = 6390$ mm

  c $y = 800$ mm, $z = 1800$ mm, $l = 10\,560$ mm, $w = 6060$ mm, $d = 4500$ mm

4 a 1 : 25 000

  b i 625 m   **ii** 1100 m or 1.1 km

  c i 3.5 cm   **ii** 10 cm

5 a 280 mm   **b** 110 mm

  c 5500 mm

  d 3500 mm by 5500 mm

  e 6390 mm by 500 mm

6 a 9

  b Length 13.1 m, width 10 m

  c 3 m

  d i 109.7 m^2

    ii \$73 719 approximately

  e 10.9%   **f** 21.3 m^2

  g i 23.43 m^2 **ii** 229 tiles

  h i 13.1 m   **ii** \$374.23

7 a \$1768

  b i 15.18 m   **ii** \$227.70

  c i 12.42 m^2 **ii** $310.5 \approx 311$ tiles

    iii $\$510.46 \approx \$511$

  d i 26 m^2 **ii** $4\frac{1}{3}$ L

    iii \$69.80 (2 cans required)

8 a No

  b i 150.4 m^2 **ii** 17 bags

    iii \$894.20

  c i 40.8 m^2 **ii** 27.1%

    iii 6.5 kW

9 Teacher to check diagram, 40 m

Chapter 13

Exercise 13.01

1 Teacher to check

2 a CD   **b** RD   **c** AD

  d PQ   **e** ICF   **f** OD

  g ICF   **h** CD   **i** PQ

  j RD   **k** ICF   **l** OD

3–5 Teacher to check

Exercise 13.02

1 4

2 3

3 Some individuals would have no relevant category available to tick; for example, a 30-year-old.

  30 years or under

  From 31 to 40 years

  From 41 to 50 years

  Over 50 years

  Other answers are possible.

4 Which of the following do you attend?

  Concerts   Dance nights

  Films   Trips away

  Other answers are possible.

5 Weekly

  Once a month

  2–3 times per month

  Over 3 times per month

  Other answers are possible.

6 For the activities you attend, the cost is:

  cheap?

  OK?

  expensive?

  I don't attend activities because they are too expensive.

  Other answers are possible.

7 Teacher to check

8 A box could be provided. Other answers are possible.

9 Teacher to check

10 Phone survey, observation, taking measurements. Other answers are possible.

Exercise 13.03

1 a Teenagers in NSW

  b All students in your school

  c Families in your street

  d Customers of pay TV

  e Customers of Cengage Bank who use Internet banking

  f Year 12 students at the school

  g People in Australia who watch TV

h People making complaints to the company

i Schools in NSW

j All people who live in the town

k People in Australia who are aged over 18 and registered to vote

l Hospitals in NSW

2 a sample   **b** sample   **c** census
d sample   **e** sample   **f** census
g sample   **h** census   **i** sample
j sample   **k** sample   **l** census

Exercise 13.04

Teacher to check

Exercise 13.05

1 a

Colour	Tally	Frequency
Black	IIII IIII I	11
Silver	III	3
White	IIII I	6
Red	IIII I	6
Other	IIII	5
		31

b

Preferred car colour

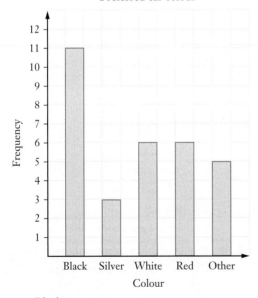

c Black

2 a

Number of phone calls per day	Tally	Frequency
1	III	3
2	IIII	5
3	IIII II	7
4	IIII	4
5	IIII	5
6	IIII	4
7	II	2
		30

b

Number of calls per day in June

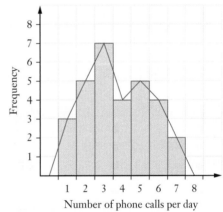

c 8 days

3 a

Number of accidents	Tally	Frequency
0	IIII I	6
1	IIII	3
2	II	2
3	II	2
4		0
5	I	1
		14

b

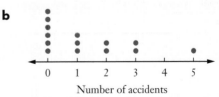

c Teacher to check

ISBN 9780170413534

4 a

Number of students served	Tally	Frequency
70–79	I	1
80–89	III	3
90–99	III	3
100–109	III	3
110–119	IIII	5
		15

b

Stem	Leaf
7	6
8	1 6 8
9	5 7 8
10	1 5 5
11	2 2 4 4 7

c Teacher to check

Exercise 13.06

1 a 12   **b** 24 minutes   **c** 141.25
   d 143   **e** 143

2 a 3   **b** 3.7   **c** 3.5
   d 50%   **e** $\frac{1}{18}$

3 a 32   **b** 20
   c 19   **d** 19

4 a 30 to 34

b For 30 scores, the median is the average of the 15th and the 16th scores. When the scores are arranged in order, the first seven scores are in the 25 to 29 range and the next 10 scores are in the 30 to 34 range. The 15th and 16th scores are both in the 30 to 34 range.

c

Class	Class centre, x	Frequency, f	$f \times x$
25 to 29	27	7	189
30 to 34	32	10	320
35 to 39	37	9	333
40 to 44	42	4	168
Totals		30	1010

d 34

5 a 30 to 39   **b** 40
   c 30 to 34   **d** $(0 + 9) \div 2 = 4.5$
   e 28.5

Exercise 13.07

1 a local garage 146c, freeway 142c
   b local garage 22c, freeway 27c
   c local garage 9c, freeway 12c
   d The local garage. Both the range and the interquartile range are smaller than at the freeway service station.

2 a 75   **b** 7   **c** 75
   d 73, 76   **e** 3   **f** $\frac{2}{15}$

3 a 30   **b** 10.1   **c** 131.5, 150
   d 18.5   **e** 6   **f** 50%
   g Yes, 50% of the scores are from the lower quartile up to the upper quartile.

4 a 15   **b** 70
   c i 25–34   **ii** 15–24
   d 24   **e** 33
   f 65   **g** 12
   h Males, the scores are more spread out

Exercise 13.08

Teacher to check

Test yourself 13

1 Teacher to check

2 a Drink what? No time frame given. Other answers possible.

b How often do you drink alcohol?

   ☐   Once a day
   ☐   Once a week
   ☐   Twice a week
   ☐   3–5 times a week
   ☐   Everyday
   ☐   Other – give details below

3 a Year 11 and 12 students at her school.

b Census – this gives the most accurate data. If the school is very large, a sample may be more appropriate for practical reasons.

4 Teacher to check

5

Score	Frequency
0	5
1	10
2	8
3	9
4	5
5	3
	40

6 a, b

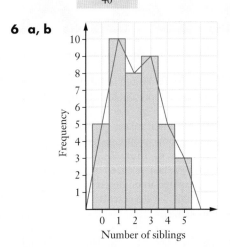

Number of siblings

7 a 2.2   **b** 2   **c** 1

8 a 5   **b** 2   **c** 1.45

9 Teacher to check

Practice set 3

Section 1

1 C	**2** D	**3** C	**4** B
5 A	**6** A	**7** B	**8** B
9 D	**10** C	**11** D	**12** A

Section 2

1 156 beats per minute

2 a $\theta = 47°27'$ **b** $x = 21.43$ m

3 a Teacher to check

   b **i** use of 'fabulous', no negative options

     ii Teacher to check

4 a Scale factor $= 3$, $x = 10.5$

   b Scale factor $= \dfrac{2}{3}$, $m = 24$

5 a 202 beats per minute

   b 160.0 beats per minute

   c 174.4 beats per minute

6 a 400 cm = 4 m   **b** 300 cm = 3 m

7 a S 53° W   **b** 233°

8 a

Score	Tally	Frequency
5–9	I	1
10–14		0
15–19	II	2
20–24	IIII	4
25–29	IIII	4
30–34	IIII III	8
35–39	IIII	4
40–44	IIII	4
45–49	III	3

b

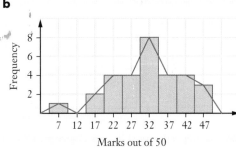

Marks out of 50

   c 11

9 a 30–34   **b** 31.7

   c 40   **d** 31

   e $Q_1 = 25$, $Q_3 = 39$, IQR $= 14$

10 a   **b** 92.1 m

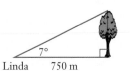

11 a 112 mmHg   **b** 77 mmHg

   c Yes

12 112.5 m

13 a 5.5 m by 5.5 m

   b **i** 30.25 m²   **ii** \$2268.75

   c 12

   d **i** 12.24 m²   **ii** \$1040.40

   e **i** 30.76 m²   **ii** \$99.80

14 a 7.425 hours of sleep   **b** 1.46

15 a 177 km   **b** 397 km

16 Teacher to check

GLOSSARY AND INDEX

adjacent side: In trigonometry, the side next to the reference angle. (p. 262)

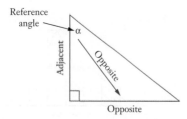

angle of depression: When an observer looks at an object that is lower, the angle between the line of sight and the horizontal. (p. 317)

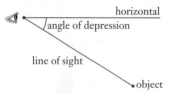

angle of elevation: When an observer looks at an object that is higher, the angle between the line of sight and the horizontal. (p. 317)

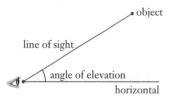

appreciation: Increase in value of an item or asset over time. (p. 28)

association: Statistical term referring to the relationship between two variables. (p. 133)

bearing: A direction from one point on the Earth's surface to another. There are two types of bearings: compass bearings and true (three-figure) bearings. (p. 333)

bias: In statistics, an unwanted influence that stops a sample from being representative of a population. (p. 396)

bivariate data: Data that relates two variables measured in the same group, such as height and weight of students. (p. 130)

blood pressure: Pressure exerted by circulating blood. It is measured in millimetres of mercury (mmHg). (p. 296) *See also* **systolic pressure** and **diastolic pressure**.

box plot (or **box-and-whisker plot**): Diagram that displays the quartiles of a set of data as a box and the extremes as whiskers. (p. 405)

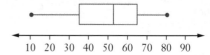

break-even point: The point or value of sales at which a business stops making a loss and starts making a profit. (p. 74)

census: Collection of information about every member of a population. (p. 392)

class centre: The centre of a class interval. For example, the class centre of the class interval 10–19 is 14.5. (p. 403)

class interval: In statistics, when there are many data scores, they may be grouped into class intervals. For example, ages of people may be grouped into class intervals of 1–10, 11–20, 21–30, and so on. (p. 403)

closed path: A path in a network that begins and ends at the same vertex. (p. 222) *See also* **open path**.

compass bearing: A bearing given as an angle either side of north and south. (p. 323) *See also* **true bearing**.

compound interest: Interest paid on the principal invested as well as on any accumulated interest. Differs from **simple interest**. (p. 10)

compounding period: How often interest is calculated when using compound interest; for example, monthly, quarterly or yearly. (p. 15)

cosine: A ratio in a right-angled triangle:

$$\cos \theta = \frac{\text{side adjacent to } \theta}{\text{hypotenuse}} \text{ where } \theta \text{ is an angle.}$$

(p. 267) *See also* **sine** and **tangent**.

declining-balance depreciation: Method of calculating depreciation using the same percentage decrease over each time period. (p. 89)

dependent variable: A variable whose value depends on another variable. It is represented on the vertical axis of a scatterplot. (p. 130)

degree (of a vertex): The number of edges that meet at the vertex. (p. 223)

depreciation: Loss in value of an item or asset over time. (p. 86)

diastolic pressure: Blood pressure in the arteries when the heart muscle is not beating (between beats). (p. 296) *See also* **blood pressure** and **systolic pressure**.

directed network: A network with arrows on the edges and movement can be in the direction of the arrows only. (p. 234) *See also* **network**.

dot plot: Graph that uses dots to show frequencies of data scores (e.g. temperatures in °C of 10 hospital patients). Different to a **stem-and-leaf plot** (p. 406)

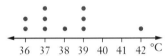

edge: In networks, a line joining vertices to each other. (p. 222)

elevation view: On a plan, the front, back or side view of a building. (p. 358)

exponential decay: A situation where a quantity decreases repeatedly by a percentage of itself. When we graph a quantity that is decaying exponentially, the graph looks like this: (p. 172)

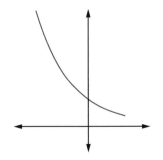

exponential growth: A situation where a quantity increases repeatedly by a percentage of itself. When we graph a quantity that is growing exponentially, the graph looks like this: (p. 172)

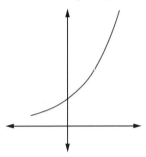

extrapolation: Making a prediction outside the range of the original data. (p. 147) *See also* **interpolation**.

frequency histogram: A bar chart in which the height of each column represents the frequency of a single score or group of scores. (p. 399)

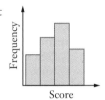

frequency polygon: A line graph formed by joining the midpoints of the tops of the columns of a frequency histogram. (p. 399)

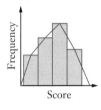

fuel consumption: Rate at which fuel is used by a vehicle, usually measured in L/100 km. (p. 212)

gradient (symbol m): Slope of a line. (p. 62)

$$\text{gradient} = \frac{\text{rise}}{\text{run}} = \frac{\text{change in } y}{\text{change in } x}$$

heart rate (or **pulse**): Speed of your heartbeat measured in beats per minute. (p. 288)

histogram: *See* **frequency histogram**.

hypotenuse: The longest side of a right-angled triangle; the side opposite the right angle. (p. 262)

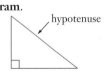

image: A figure that is an enlargement or reduction of an original figure. *See also* **scale factor**. (p. 348)

independent variable: A variable whose value does not depend on another variable. However, the value of the dependent variable depends upon the independent variable. It is represented on the horizontal axis of a scatterplot. (p. 130) *See also* **dependent variable**.

inflation: Rate at which the overall cost of goods and services is increasing. (p. 26)

interpolation: Making a prediction within the range of the original data. (p. 147) *See also* **extrapolation**.

interquartile range (IQR): Difference between the upper quartile and the lower quartile of a data set $(Q_3 - Q_1)$. It is a measure of the spread of the data. (p. 404)

Kruskal's algorithm: Method of finding a minimum spanning tree for a weighted network. (p. 246) *See also* **Prim's algorithm**.

line of best fit: A straight line that represents a set of points on a scatterplot, obtained through experiment or observation (p. 139)

linear function: A function of the form $y = mx + c$, the graph of which is a straight line. (p. 62)

linear modelling: Using a linear function to approximate a real-life situation. (p. 66)

lower quartile: is the 1st quartile $(Q1)$ that cuts off the bottom 25% of scores in a data set. (p. 404) *See also* **quartile**.

map scale: *See* **scale**.

mean: The average of a set of scores. (p. 399)

$$\text{mean (or } \overline{x}) = \frac{\text{sum of scores}}{\text{number of scores}} = \frac{\sum x}{n} = \frac{\sum fx}{\sum f}$$

measure of central tendency: A statistical value such as the mean, median or mode that describes the centre or average of a set of data. (p. 399)

measure of spread: A statistical value such as the range, interquartile range or standard deviation that describes the spread of a set of data. (p. 404)

median: The middle score of a data set when scores are arranged in ascending order. If there are two middle scores, the median is the average of the two. (p. 399)

minimum spanning tree: A spanning tree of minimum length for a network. (p. 243) *See also* **spanning tree**.

mode: The most common or frequent score(s) in a set of data. (p. 399)

network: Diagram comprising vertices and edges representing a system of interconnecting objects; e.g. a bus network. (p. 222)

non-linear function: A function whose graph is a curve, *not* a straight line. (p. 166)

open path: A path in a network that begins and ends at different vertices. *See also* **closed path**. (p. 222)

opposite side: In trigonometry, the side facing the reference angle in a right-angled triangle. (p. 262)

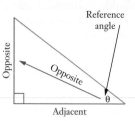

parabola: U-shaped graph of a quadratic function. (p. 169) *See also* **quadratic function**.

path: A route in a network where there are no repeated edges or vertices. (p. 222)

plan: Diagram of the floor of a house or building. (p. 358)

population: In statistics, all of the items under investigation. (pp. 395, 396) *See also* **census** and **sample**.

Prim's algorithm: A method of finding a minimum spanning tree for a weighted network. (p. 243) *See also* **Kruskal's algorithm**.

principal: The original amount of money invested or borrowed. (p. 6)

Pythagoras' theorem: In a right-angled triangle, the square of the hypotenuse is equal to the sum of the squares of the two shorter sides. $c^2 = a^2 + b^2$. (p. 258)

quadratic function: A function of the form $y = ax^2 + bx + c$. Its graph is a **parabola**. (p. 169)

quartiles: Values that divide a set of data into four equal parts when the scores are arranged in order. The 1st quartile (Q_1) is the lower quartile, the 2nd quartile (Q_2) is the median and the 3rd quartile (Q_3) is the upper quartile. Q_3 cuts off the top 25% of scores in a data set. Q_1 cuts off the bottom 25% of scores. (p. 404)

range: Difference between the highest score and the lowest score in a set of data. (p. 404)

rate: A comparison between two quantities of different types; for example, kilometres per hour. Different to **ratio**. (p. 192)

ratio: A comparison between two quantities of the same type; for example, 4 : 7. Different to **rate**. (p. 36)

reducible interest: Paying interest only on the amount of money still owing, not on the original amount borrowed. As the amount owed decreases, the interest payments decrease. (p. 92)

reducing balance loan: A reducible-interest loan where the interest charged is calculated on the balance owing on the loan after each repayment. (p. 93)

salvage value: Value of a depreciating item. (p. 86)

ISBN 9780170413534

sample: A group of items selected from a population. (p. 392)

scale (on a map or diagram): The ratio of scaled length to actual length; for example, a scale of 1 : 500 means that lengths represented on the map or diagram are actually 500 times larger in real life. (p. 49)

scale drawing: A drawing of an object, usually smaller, whose lengths are in the same ratio as the actual lengths of the object. (p. 354)

scale factor: The ratio $\dfrac{\text{image length}}{\text{object length}}$ for a pair of similar figures. For example, a scale factor of 3 means that every length of the image is 3 times the corresponding length of the object. *See also* **image**. (p. 348)

scatterplot: A graph of points on a number plane showing a relationship between two variables. (p. 130)

shortest path: The path in a weighted network where the sum of the edges is minimised. (p. 226)

similar figures: Figures that have the same shape but not necessarily the same size. Matching sides are in the same ratio. Matching angles are the same. (p. 348)

simple interest (or **flat-rate interest**): Interest earned or charged only on the original amount of money (principal) invested or borrowed. Differs from **compound interest**. (pp. 6, 92)

simultaneous equations: A pair of equations, such as $y = x + 2$ and $y = 2x - 3$, that can be solved together to find the values of x and y that satisfy *both* equations. On a graph, the solution is represented by the point of intersection of the graphs of the equations. (p. 70)

sine: A ratio in a right-angled triangle:

$\sin \theta = \dfrac{\text{side opposite to } \theta}{\text{hypotenuse}}$ where θ is an angle. (p. 267)

See also **cosine** and **tangent**.

site plan: A drawing showing where a building is sited on a block of land. (p. 373)

spanning tree: A tree in a network that connects all of the vertices of the network. A network can have more than one spanning tree. (p. 239) *See also* **tree** and **minimum spanning tree**.

standard deviation (symbol σ): A statistical measure of the spread of a set of scores. (p. 404)

stem-and-leaf plot: A 'number graph' that lists all the data scores in groups. This stem-and-leaf plot shows 12 test scores from 42 to 82. Different to a **dot plot**. (p. 400)

Stem	Leaf
4	2 5
5	0 2 8
6	6 7
7	3 5 7 7
8	2

straight-line depreciation: Method of depreciation in which an item's value decreases by the same amount each period. (p. 86)

systolic pressure: Blood pressure in the arteries when the heart muscle is beating. (p. 296) *See also* **blood pressure** and **diastolic pressure**.

tangent: A ratio in a right-angled triangle:

$\tan \theta = \dfrac{\text{side opposite to } \theta}{\text{side adjacent to } \theta}$ where θ is an angle.

(p. 262) *See also* **sine** and **cosine**.

target heart rate: Optimum heart rate during aerobic (cardiovascular) training, providing maximum health for the heart and lungs. (p. 292)

target population: The particular subgroup used for a statistical investigation; for example, women under 40 years of age. (p. 395)

tree: A network in which any two vertices are connected by exactly one path. It does not contain any cycles where you can return to the same vertex along a different edge. *See also* **spanning tree**. (p. 239)

true bearing (or **three-figure bearing**): Written as a 3-digit angle from north going in a clockwise direction, from 000° to 360°. (p. 328) *See also* **compass bearing**.

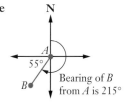

Bearing of B from A is 215°

unit price: Giving the price for a standard amount of an item; for example, showing the cost for 100 g of different types of margarine. Unit pricing allows comparison between different items. (p. 206)

unitary method: A way of calculating a value by first finding one unit and then multiplying. (p. 40)

upper quartile: The 3rd quartile (Q_3) that cuts off the top 25% of scores in a data set. (p. 404) *See also* **quartile**.

vertex (plural: vertices): A point in a network where edges meet. (p. 222)

vertical intercept: *See* **y-intercept**.

weighted edge: An edge in a network that has a number on it. The number could represent distance, time, cost or other variables. (p. 222)

y-intercept (or **vertical intercept**): The value at which a straight-line graph cuts the *y*-axis. For example, the *y*-intercept of this graph is 3. (p. 62)

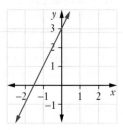